"VI GRAVITATIS"

New Theory of Universal Gravitation.
Uniform and eccentric circular motion of the
planets and calculation of their orbits

Title:
Vi Gravitatis

Subtitle:
New Theory of Universal Gravitation. Uniform and eccentric circular motion of the planets and calculation of their orbits

Autor:
Benito Vinuesa Guerrero

Copyright
All rights reserved.

ISBN: 9781081459765
Sello: Independently published

Back cover:
Photo of the author in the yard of the *Carmen de los Mártires*, next to the Alhambra in Granada, Spain.

"On a trembling point will be Gebenite law"

3rd verse; Quatrain LXIV; Centurie II; Nostradamus

"A Palmira, Natalia y Miguel Ángel"

Granada, (Spain); July 20, 2019

About the Author:

The author, born in Alhama de Granada, (Spain), in 1957, has a degree in Physics and a PhD from the University of Granada. He is also an engineer and a Doctor engineer from the Higher Polytechnic School of the Army, Madrid.

We appreciate the suggestions and comments of the readers, on the subject treated and on what is related to the errors and deficiencies of this edition. Your help is important because with it you can improve this book and advance knowledge. Feel free to communicate and send your comments to this email address: bvin01@gmail.com

Index

1. Introduction..7
2. Summary...11
3. New Theory of Universal Gravitation................................13
   3.1 Background. Newton's Law of Gravitation.....................13
   3.2 New theory proposed...14
4. Proposed validation experiments.....................................25
   4.1 Hydrogen balance..25
   4.2 Modification of the Cavendish balance.........................27
   4.3 Measure of pendulum inclination next to a large cliff......28
5. Uniform and eccentric circular motion of the planets..........29
   5.1 Background. Geocentric and heliocentric models...........29
   5.2 Kepler's laws..31
   5.3 Proposed model..31
      5.3.1 Relationship between position, speed and acceleration
     41
   5.4 Comparison of Kepler's laws with the new model..........44
6. Calculation of the orbits of the Moon and the planets.........49
   6.1 Integration of the new acceleration equation................49
   6.2 Equation of periodic movement.................................51
   6.3 Astronomical data used for the calculation..................53
   6.4 Orbit of the Moon..61

    6.5 Orbits of the planets..67

        6.5.1 Mercury Orbit..69

        6.5.2 Orbit of Venus...74

        6.5.3 Earth orbit..78

        6.5.4 Mars Orbit..83

        6.5.5 Orbit of Ceres..89

        6.5.6 Jupiter's orbit...95

        6.5.7 Saturn's orbit..101

        6.5.8 Orbit of Uranus...107

        6.5.9 Orbit of Neptun...113

7. Proposed calculation of the equilibrium distance..................117

8. Impact of asteroids on Earth..123

9. Annex: Spreadsheet formulas..125

    9.1 Mercury orbit formulas..126

1. Introduction

Just on July 20, 2019, fifty years of a planetary event are celebrated. The man stepped on the moon. The term is now fulfilled in which the great Brazilian popularizer and famous medium Francisco Cándido Xavier, popularly known as Chico Xavier, expected a second planetary event in the Solar System. This event would lead humanity to a new era of Peace and Prosperity for humanity and planet Earth. That event, which we also want to happen, seems to be coming late. We, on the other hand, have wanted to be punctual and take advantage of the date to give birth to this book.

Coinciding with this 50th anniversary, this New Theory of Universal Gravitation is presented. It includes the repulsive term that is missing in Newton's equation. By integrating this new equation, the orbits of the Moon and the planets are obtained with simple calculations. It turns out that these orbits are eccentric circles traversed by them with uniform circular motion.

With the cover making a nod to the famous book

"*Philosophiae Naturalis Principia Mathematica*" by Sir Isaac Newton, we wanted to pay tribute to him. Its gravitational law has ruled since 1687. Three hundred and thirty-two years! Our recognition and admiration also to all the men and women who have dedicated their time and effort to the development of Science. On the subject of gravity and celestial orbits, six wise men have stood out throughout history. We refer to Aristotle, Claudius Ptolemy, Nicolaus Copernicus, Galileo Galilei, Johannes Kepler and Isaac Newton. Following in his footsteps we have reached where we are today.

The present theory, the result of many calculations and reflections, part of current knowledge and has been possible thanks to the help of the Internet and the calculation capacity of current computers. It aims to be a contribution to the great collective building of science and one more step in the knowledge of the world that surrounds us.

This book has had a slow and difficult gestation. It has been published when the author, early retirement by age, has been able to get enough free time. They have been fringed but is made known so that other researchers can collaborate in its validation and because there are currently underway studies and projects to bomb asteroids and thus supposedly prevent them from impacting the Earth. This in the light of the new theory, if you do not take into account the mass relationship, can be a real nonsense. It is important that this theory is known and validated.

Thanks to the effort and sacrifice of my parents and the dedication of my teachers I was able to study and acquire the necessary knowledge to write this book. For them my deepest gratitude.

Note to this English edition:

The author expresses his gratitude to the Google translator. Thanks to him, this book will be accessible to readers from all over the world who understand English, the language of the great physicist Isaac Newton.

The author asks readers for understanding because some table texts could not be translated and apologizes for possible mistakes.

2. Summary

Newton's Law of Gravitation, although functioning on the surface of the Earth, can not be universal or applicable to the movement of the planets. It is just attraction, so it is intrinsically unstable and if true, the universe would have collapsed long ago. Based on two postulates, a New Law of Universal Gravitation is developed here, of which Newton's law is a particular case. This New Law has an attractive and repulsive term dominating one or the other according to whether the distance to which the bodies are located is greater or less than an equilibrium distance do. The New Law is:

$$F = G M m \left(\frac{1}{d^2} - \frac{1}{d_0^2} \right)$$

This New Law produces an indefinite oscillatory movement within a stable equilibrium.

With this New Law the orbits of the Moon and the planets are calculated very easily. When integrating the equation, it turns out that the orbit of them is an eccentric circumference tangent to two other circles centered on the Sun with radii Aphelion, A and Perihelion, P respectively. The orbit, highlighted in the figure, is traversed with a uniform circular motion. Kepler's first law, although approximate, turns out to be untrue. Yes it is true the second and the third comes out very similar but not exact. It turns out that the radius of the orbit is the semi-summit of the apses, the decentered of the Sun the semidifference and the distance of equilibrium the geometric mean. It can not be much simple!

$$R=\frac{(A+P)}{2} \quad ; \quad D=\frac{(A-P)}{2} \quad ; \quad E=\sqrt{AP}$$

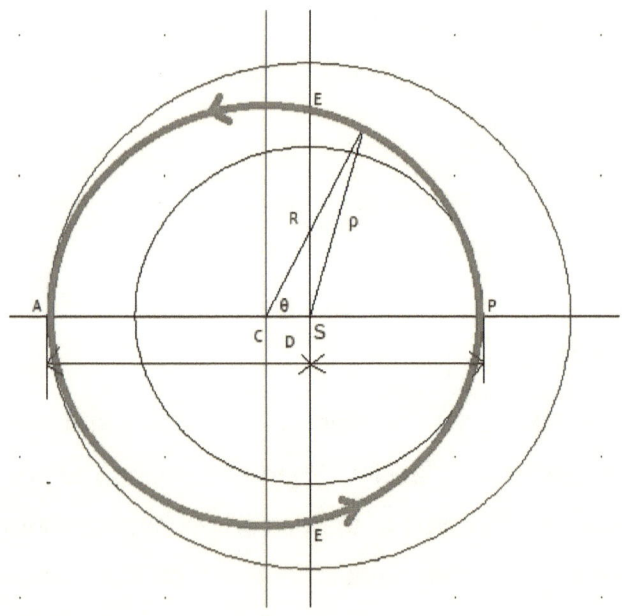

3. New Theory of Universal Gravitation

3.1 Background. Newton's Law of Gravitation

Isaac Newton in 1687, after almost twenty years of waiting, published his work: *"Philosophiae naturalis principia mathematica"*. This work is considered by many to be the culmination of the scientific revolution initiated by Copernicus and followed by Galileo and Kepler. In it is contained his law of universal gravitation. It is said that he deduced it from Kepler's laws. The statement of this law is as follows: given any two bodies of masses M and m, separated by a distance d, the force of attraction exerted between them is equal to:

$$F = \frac{-G M m}{d^2}$$

The minus symbol is because force is always attractive, that is, tends to decrease d. G is an equal universal constant for all bodies called the universal constant constant of gravitation. Vector notation has not been used for simplicity. Force and distance are vectors in this equation.

3.2 New theory proposed

A New Law of Universal Gravitation is proposed. The ignorance of the internal processes that intervene in the phenomenon of gravitation and the difficulty to demonstrate the real existence of gravitons leads us to deduce it planetting from two postulates. These particles are so small that with current instruments they are neither measurable nor observable. They are only detectable by their gravitational effects. Admitting its existence is more coherent than admitting mysterious forces at a distance that interact with matter without being matter.

First postulate: gravitational forces originate as a consequence of impulse transfer by physical contact through very small particles that move at the speed of light. These particles are continuously emitted by matter in all directions in proportion to their mass.

Second postulate: The flow of gravitons from a body M when it reaches another body m, cancels the same number of gravitons emitted by m in the direction of M. The effect of this cancellation is that in the direction that joins the two masses, m is emitting less gravitons in the direction towards

M and more in the opposite direction, thus producing a reaction force that drives am toward M. The same happens when the particle flow of m reaches M.

From these two postulates a New Law of Universal Gravitation is immediately deduced without needing to resort to a mysterious force that magically exerts its action at a distance without any contact and without intervention of any matter.

The gravitational interaction between two bodies that for simplicity are assumed to be spherical of radii R and r and of masses M and m respectively with M > m is the following. Suppose initially that both bodies are separated by a very large distance d. Both bodies are emitting a number of gravitons proportional to their mass. Respectively: kM and km. Since the emission is radial and in all directions, the number of gravitons that reaches any surface is inversely proportional to the square of the distance at which that surface is located. So the number of them coming from M that reaches the circular section of m is equal to:

$$\frac{kM\pi r^2}{4\pi d^2} \quad (1)$$

initially d is taken large enough so that (1) is less than the amount that emits m in that direction. Following the second postulate, these beams cancel each other proportionally, leaving in one direction or another the unbalanced balance of gravitons. The balance for the body m, also for the M, is that in the direction towards M, it is emitting less gravitons than in the opposite direction. Just the ones that he receives from M. By reaction a force is produced towards M that corresponds to the number of received gravitons. The force of interaction is proportional to the mass of m:

$$F = \frac{-GMm}{d^2} \quad (2)$$

It is negative because it is attractive, that is, it tends to decrease d. This is the well-known force of attraction of Newton that as it is seen is of reaction. As m approaches M decreasing d, the number of gravitons that reach m increases with the reaction force increasing in the same proportion. The maximum of the reaction force fmax will be reached when the number of gravitons received is equal to that which emits m in that direction. At the distance where this occurs, we call it dmax. This distance depends on the radius and the mass of both bodies. From here Newton's gravitational law is no longer fulfilled. In other words, the previous law is only applicable when the bodies are at a distance d> dmax. This applies to bodies on the surface of the Earth but not to the Moon and the planets as will be seen later.

So you have to:

$$f_{max} = \frac{GMm}{d_{max}^2} \quad (3)$$

The reaction force can not be greater because in reality it comes from m and at this distance the maximum of its emission has been compensated. This force is attractive and makes the distance d continue to shorten. The excess of received gravitons produces a force of push that little by little compensates the reaction force causing it to decrease until it is canceled when the received gravitons coincide with twice those emitted in that direction. It is double because one part cancels those that go in that direction and the other half cancels those in the opposite direction that produce the reaction force. At that distance we call it balance distance do. When the bodies are separated their equilibrium distance the force between them is null.

$$-2f_{max}+\frac{GMm}{d^2} \quad (4)$$

For smaller distances the force is clearly repulsive. The number of gravitons emitted by M that reach m is greater than twice that of emitted by m and there is no way to compensate or counteract so the radiation pressure causes m to move away from M. This equilibrium is stable but dynamic. When the body m passes the equilibrium point, the acceleration is canceled but it carries a speed that is non-zero and the inertia causes it to continue its movement until the force, which is contrary to its speed, causes the speed to be canceled and change direction and so on indefinitely if no other force acts. This oscillating movement of approach and distance is what explains the movement of the Moon and the planets since it is synchronized with the one of translation.

For distances between dmax and d the force is still attractive although less than fmax, its value is equal to:

$$\frac{GMm}{d_0^2}=2f_{max} \quad (5)$$

It decreases to zero in d = d₀ and for smaller distances it becomes repulsive.

From equations (3) and (5) it follows that:

$$d_{max} = \sqrt{2}\, d_0$$

Taking into account the signs, positive = repulsive and negative = attractive, the force of gravity for any d ≤ dmax is valid:

$$F = \frac{GMm}{d_0^2} - 2f_{max} = GMm\left(\frac{1}{d^2} - \frac{1}{d_0^2}\right)$$

When the distance is such that d> d₀ then the force becomes attractive again producing a continuous oscillatory movement of approach and distance from d₀. As stated above, this is the mechanism that explains the movement of the planets. On the Moon and the rest of the planets of the solar system, the force of attraction is such that the separation distance around d₀ never reaches, not even approaches dmax, so the law that governs all is:

$$F = GMm\left(\frac{1}{d^2} - \frac{1}{d_0^2}\right)$$

We call this law orbital law to distinguish it from Newton's.

d₀ has the property of being the geometric mean between the distances of maximum and minimum distance. For example in the case of the Sun and the Earth, aphelion and perihelion, according to the astronomical tables, are respectively:

A = 152 098 232 000 m; P = 147 098 290 000 m

then $d_0 = \sqrt{AP}$ = 149 577 370 746 m

$d_{max} = \sqrt{2} d_0$ = 211 534 346 333 m

how it looks very far from the aphelion. Since the Sun does not let escape to Earth and never reaches a distance greater than dmax, Newton's gravitational law is not applicable to the Earth's orbit around the Sun nor to any other planet. In all of them, the orbital law with a repulsive and attractive term is applicable.

When the Earth is in the aphelion it suffers attraction from the Sun and in the perihelion it suffers repulsion. In d_0 there is neither attraction nor repulsion, that is, both are compensated and the resultant force is zero.

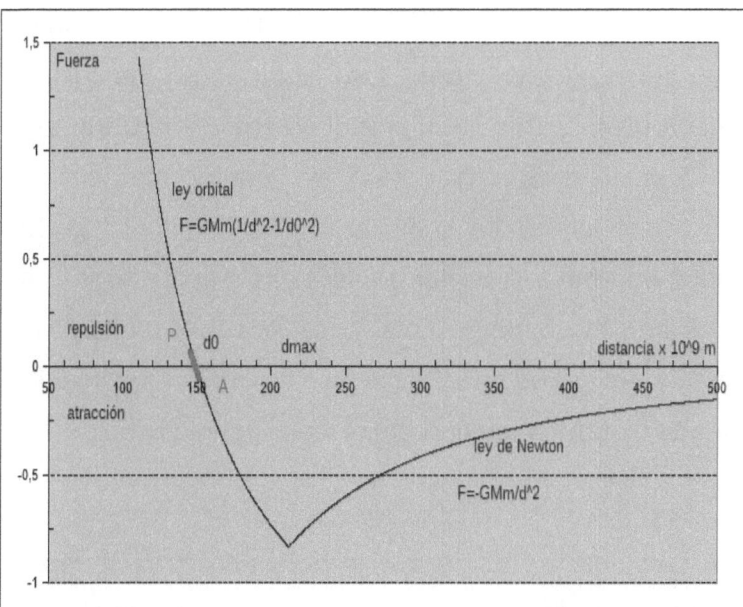

Gravitational force scheme in the Sun-Earth system

In the previous figure the gravitational force is represented in the Sun-Earth system. The Earth is trapped in an oscillatory movement moving along the line highlighted in red from P, perihelion, to A, aphelion, always to one side and another to do. The Sun is at the origin of coordinates. The positive y axis represents repulsive force and the negative part represents the attractive force. You can also see dmax from which Newton's law is valid although in this case the distance is unattainable by the Earth and therefore can not be applied. The oscillatory movement of approaching-away from the Earth to the Sun, is synchronized with a circular movement of the Earth around the Sun. The Sun is not at the center of this circular movement as will be seen later. The fact that the Sun is displaced from the center could make Kepler think that the Earth and the other planets follow elliptical orbits. It is not like this. As will be checked later. The orbits are circular. It is curious that the ancient Greeks were right in this.

The fact that the same law is both attractive or repulsive depending on the distance d is explained very easily.

$$F = GMm\left(\frac{1}{d^2} - \frac{1}{d_0^2}\right)$$

The result is that:

- For $d = d_0$ the second member becomes equal to zero and the force is canceled.

- For $d < d_0$ when both are in the denominator, the force is positive. F and d have the same sign ie when F increases, d increases and this is a repulsive force.

- For $d > d_0$ when both are in the denominator, the

force is negative. F and d have a different sign, that is when F increases, d decreases and this is an attractive force.

The dynamic equilibrium given by this law is a stable equilibrium. The concept of stability for a balance is very important and has to do with the return or not to balance after a disturbance that separates the body from equilibrium. If when leaving the equilibrium position the body remains as it is, the balance is indifferent. If there is a force that causes it to separate more, the balance is unstable and if that force causes it to return to the equilibrium position, it will be stable. A simple example of these three types of balance is a ball on a horizontal table, a ball on top of a circular bowl that is face down on the table and a ball inside that same circular bowl but this time placed face up on the table. In the case of the new law of gravitation, the equilibrium point is d_o. If d becomes greater than d_o, then the attractive force causes d to decrease, whereas if d becomes less than d_0, the repulsive force causes d to increase. In both cases d approaches d_0, so the equilibrium is stable. Observe how Newton's law of gravitation is unstable. Assuming a body in equilibrium in orbit, any small disturbance away from the body would cause d to increase so that the attractive force would decrease in proportion to the square of the distance away, ie if the disturbing force is constant, it would move it further and further away. On the contrary, a disturbance in the sense that d decreases, as the attractive force would increase in the proportion of the distance squared, would add up and cause it to come closer still falling irretrievably.

The proposed new law of universal gravitation implies an oscillatory movement around the equilibrium point d_0. That is what is observed with the Moon around the Earth and of this and the other planets around the Sun. With the galaxies and with everything the same thing happens. The fact that the galaxies are moving away at this moment only indicates that the distance they are currently at is less than d_0, their equilibrium distance which in this case is very large given the value of the masses involved. After this distance they will come closer without a doubt. It's like a systole and diastole of the Universe. The periods of these oscillatory movements are proportional to the value of the equilibrium distance. For the Earth one year, for the galaxies it is unknown but it must be enormous.

4. Proposed validation experiments

The fact that this new theory explains the planetary movement in a simple, satisfactory way and without contradictions is a clear indication of its validity. The values of the constants and individual masses remain to be determined, not of their product. There are other forms of experimental validation. The author of this theory does not have the means to carry them out or to obtain the value of these constants. That is why several experiments are proposed to the scientific community and to any interested person who has the means to carry them out.

4.1 Hydrogen balance

We call hydrogen balance a simple device that allows you to check the theory proposed in any workshop or laboratory. It consists of a tube closed by the ends of several meters in length and several centimeters in diameter. Made with resistant material so that the vacuum can be made inside and not deformed. The tube must be supported at its center and be able to rotate to become vertical, being able to place either end at the top. On the inside of each end you must have a sensor to measure the concentration of hydrogen.

The tube is evacuated and then a small amount of hydrogen is introduced through a suitable valve. When the tube is horizontal, after a time to reach equilibrium, the sensors must measure the same concentration of hydrogen. The experiment consists of measuring the concentration of hydrogen at each end when the tube is vertical after a time so that the gas reaches equilibrium with both ends at the same temperature. Within the tube there is only hydrogen so it does not influence the principle of Archimedes. With the vertical tube at rest, the hydrogen inside the tube is only subjected to an external force that is gravity. According to Newton's classical theory, the lower sensor must register a higher concentration of hydrogen. If we add the gravitational attraction to the random movement of the hydrogen molecules, there must be a higher concentration below.

According to the proposed new theory, the equilibrium distance x_0 for hydrogen is greater than the terrestrial radius, so gravity would act by repelling instead of attracting. That is, there will be a higher concentration of hydrogen in the upper end of the tube. This same experiment can be done with other gases that have x_0 greater than the height where the tube is located. Hydrogen has been proposed because it would be more sensitive to equal pipe length. Everything indicates that the terrestrial atmosphere has already carried out this experiment but it is necessary to repeat it without the presence of other gases to rule out the push of Archimedes by other gases with greater density.

This same experiment conducted in laboratories located at different distances from the center of the Earth could give values of the gravitational force of repulsion. It would be a way of weighing the Earth by measuring the repulsion of the hydrogen molecules, hence the name we have given it of the hydrogen balance.

4.2 Modification of the Cavendish balance

The Cavendish balance is a very precise instrument that has been used to calculate the value of G, the universal gravitation constant. This has been perfected since in 1798 Henry Cavendish managed to measure in a laboratory the attraction between two known masses. A modification of the instrument is proposed in order to detect the existence of the repulsive term of the new gravitational law. The proposed modification consists in that the fixed and mobile masses are enlarged and decreased respectively so that their values are as far as possible and thus their quotient is very large. We must ensure that the equilibrium distance is greater than the separation of the masses when they are as close as possible. If this is achieved, the repulsion can be observed and measured. If the frictional forces are also reduced sufficiently, the small mobile mass will be repelled and attracted by the large fixed mass in an undefined oscillatory movement. In addition, a perpetuum mobile will be obtained in a laboratory. This is already done by the solar system on a large scale with the planetary movement.

4.3 Measure of pendulum inclination next to a large cliff

As Newton did in his day, it is proposed to measure the deviation of the plumb line from the vertical near a mountain. This was done with imprecise results by several scientists, the first of them was the Frenchman Pierre Bouguer with his team in the year 1738 who made the measurements at the *Chimborazo* volcano, a great mountain of the Andes mountain range in Ecuador. Later in 1776 a team of scientists from the Royal Society of London headed by Maskelyne repeated the experiment on Mount Schiehallion in Scotland.

On this occasion, what is intended is to observe and measure the repulsive term of the new gravitational law, so the distance from the pendulum to the great mass must be as small as possible. It is proposed to repeat the experiment on a cliff that meets the appropriate conditions.

5. Uniform and eccentric circular motion of the planets

5.1 Background. Geocentric and heliocentric models

The geocentric model is due to Claudio Ptolemy, astronomer and astrologer who worked in the Library of Alexandria in the second century of our era. He devoted himself to astronomical observation and his work *"Almagesto"* contains a catalog of all the planets known in antiquity. It also includes the teachings of Hipparchus whose writings have not reached us. He compiled all the available astronomical data and made a geometric model of the movement of the planets. This model has been the reference until the sixteenth century. The observations led him to eccentric orbits. His model explained the planetary movements and could even predict eclipses. It was a very important breakthrough in introducing empiricism and observation. The main difficulty, later overcome by Copernicus, was that his model was geocentric, which excessively complicated the description of the movements by having to establish, among other things, epicycles and retrograde movements in order to explain the data. For him the Earth was fixed and immobile and was the center of the universe.

Nicolaus Copernicus, a Polish astronomer who lived from 1473 to 1543, is considered the father of the heliocentric theory. This theory was already defended in ancient times by the Pythagoreans and by Ariplanetchus of Samos. This does not detract from Copernicus who corrected the current geocentric model. This in his time was an authentic revolution. In addition to being an astronomer, Copernicus, who was a Catholic clergyman with many occupations and responsibilities, took almost twenty-five years to develop his model. This was published posthumously the same year of his death. His work: *"De revolutionibus orbium coelestium"* is considered the beginning of modern astronomy. Galileo Galilei, professor of mathematics, physicist and Italian astronomer who lived from 1564 to 1642, defended the heliocentric model of Copernicus. His controversy with the Catholic Church on this subject is famous.

Johannes Kepler, German astronomer and mathematician who lived from 1571 to 1630, perfected the heliocentric model. From the observations of the Swedish astronomer Tycho Brahe deduced his three laws of the planetary movement around the Sun. He is considered the best astronomer of his time and his excellent work was a very important advance in astronomy. His three laws of motion together with Newton's gravitational theory are the current basis for the current calculation of planetary movements.

5.2 Kepler's laws

- **First:** All the planets move around the Sun following elliptical orbits. The Sun is in one of the foci of the ellipse.

- **Second:** The planets move with constant areolar velocity. That is, the vector position ρ of each planet with respect to the Sun sweeps equal areas in equal times.

- **Third:** It is true that for all planets, the ratio between the period of revolution to the square and the semimajor axis of the ellipse to the cube remains constant. This is:

$$\frac{T^2}{a^3} = k$$

5.3 Proposed model

By applying the new law of universal gravitation to the movement of the Moon and the planets, bearing in mind that this accelerated oscillatory movement of attraction repulsion is synchronized perfectly with an orbital movement of the planet around the Sun or the Moon around the Earth. The orbits are perfectly circular and both the planets and the Moon move in them with uniform circular motion. Both the Sun and the Earth are off center.

The previous results have arisen by applying the appropriate gravitational law. At first it was expected to obtain elliptical orbits according to Kepler's laws. The calculations have been revised and circumferences always come out. To be sure of this, to the spreadsheet apart from the radius vector ρ from the Sun to the planets, the velocity and acceleration of the attraction-repusion movement, three new columns have been added, a new circular ρ, the radius R calculated from the law of cosines and the arrow or deviation from the radius of the perfect circumference. As the iteration time increment decreases, that is, as the number of rows in the spreadsheet increases, the arrow decreases. Initially the angle θ had the value of one degree each row, there were 360 rows. It has been passed to 3600, 36000 and finally to 360000 rows that is the maximum that our computer admits without being blocked. The arrow has been decreasing each time.

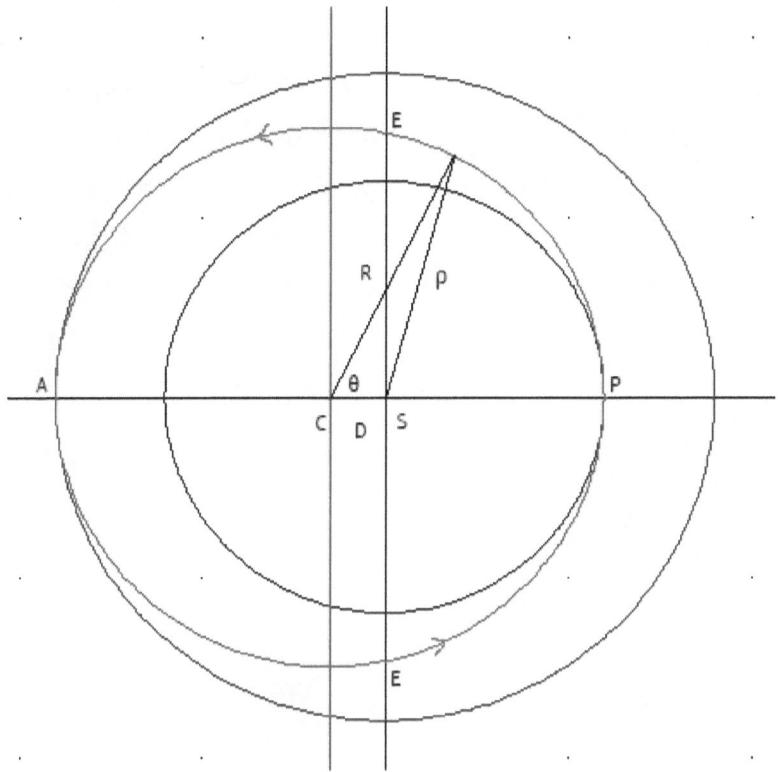

Synchronized circular orbit

The body with the highest mass, in this case the Sun, is assumed to be fixed at point S. Around this point two concentric circles of radii P and A respectively are traced. They are the perihelion and the aphelion. A third tangent circle is drawn on the outside to the one with radius P and tangent on the inside to the one with radius A. This third circumference is the orbit of the planets. The Moon and the planets, in addition to the rotation on their axis, have two perfectly synchronized movements. One of separation approach to the Earth or the Sun as appropriate and another circular around the center C. The synchronization is that the time it takes to make a complete turn T, sidereal period, is the same time it takes to get away and Go back to the planetting point. The uniform circular motion has a radius R that is the arithmetic mean between apogee and perigee.

$$R = \frac{(A+P)}{2}$$

The angular velocity and the linear velocity are constant and are given by:

$$w=\frac{2\pi}{T} \quad ; \quad v=wR$$

The Earth, in the case of the Moon, or the Sun in the case of the planets, are displaced from the center of the C orbit, a distance D that is the semidifference between the apogee or aphelion and the perigee or perihelion respectively in each case.

$$D=\frac{(A-P)}{2}$$

This displacement is what could have induced Kepler to think that the orbits were elliptical. You have to think that the distances are measured from the Earth and the radius vector ρ varies.

The other movement synchronized with this, is that of distance approach given by the law of gravitation, is an oscillatory movement of variable acceleration with the square of the distance. Its acceleration is:

$$\ddot{\rho}=k\left(\frac{1}{\rho^2}-\frac{1}{E^2}\right)$$

ρ is the radius vector measured from the Earth, E is the

equilibrium distance d_0 which is the geometric mean of A and P.

$$E=\sqrt[2]{AP}=\sqrt[2]{R^2-D^2}$$

Coincides that E is the straight distance up and down the cut of the orbit with the vertical drawn by the Earth or the Sun in its case.

Interestingly, if instead of a circumference, the orbit were an ellipse, E would be *the semilatus rectum* of the ellipse. This seems a coincidence but it is given by being the geometric mean of A and P. On the other hand it is curious to note that given the aphelion and perihelion, or apogee and perigee, A and P, the parameters of the orbit are given by their mean arithmetic, its semidifference and its geometric mean. Three different averages!

While $\rho <E$ the body of greater mass, the Earth or the Sun in the case of the planets, exerts a repulsive force on the body of smaller mass. When $\rho > E$ the force becomes attractive although the body of smaller mass m keeps moving away by inertia a distance whose maximum is A, it is attracted each time with less force until the attractive force is annulled in $\rho = E$ and from there it becomes repulsive but also the body continues to approach by inertia to a minimum distance given by P. The body of smaller mass is trapped and continues indefinitely circular movement around C, which if considered with respect to the Earth or the Sun, is of approaching distance. It is important to note that this movement is stable, always returns to the point of equilibrium. When the body of smaller mass approaches, it is repelled by the body of greater mass and when it moves away it is attracted. Newton's law in this sense is quite the opposite, it is completely unstable. Any small disturbance that caused it to move away or get closer, would imply increasing distance or approach indefinitely, never return to equilibrium.

The relation between the radius of the orbit R and the radius vector ρ, is given by the cosine theorem.

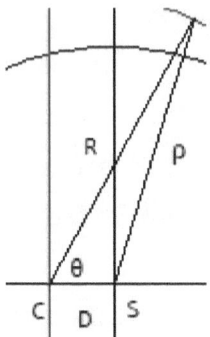

Cosine theorem

Applying the cosine theorem to the triangle of sides R, ρ and D, we have to:

$$\rho^2 = R^2 + D^2 - 2RD\cos(\theta)$$

$$\rho = \sqrt{R^2 + D^2 - 2RD\cos(\theta)}$$

$$\rho = \sqrt{\frac{A^2 + P^2 - (A^2 - P^2)\cos(\theta)}{2}}$$

The accelerated oscillatory movement of attraction-repulsion originated by the gravitational force is perfectly synchronized with the circular orbital movement. Beginning to count in the perihelion with 0 degrees, in a half turn, at 180 degrees the planet is in the aphelion and in the complete turn, that is to say 360 degrees sexagesimal, it is found again in the perihelion.

At x_0, equilibrium distance, we have called it E and it is somewhat before 90 degrees when it goes from perihelion to aphelion and somewhat after 270 degrees when it goes from aphelion to perihelion. The variation depends on the separation distance between the Sun and the center of the orbit. Kepler searched for the circumference but could not find it because he took the radio vector as a reference. You have to make a change of coordinates by putting R in the center and of course take the angles θ from this center and not from the radius vector ρ that goes from the Sun to the planets. The astronomical observations give the latter. Kepler tried to make a change of angles introducing eccentric, average and true anomalies but with the means he had he could not see the circumference. In his honor it must be said that doing it in this way from observations like this is very difficult. We have done it the other way around planetting from the gravitational equation. The circular orbit has gone out alone by integrating the equation of the appropriate law of gravitation. The computer and the spreadsheet, of which Kepler did not have, have facilitated these calculations that give like result a circular orbit and a uniform orbital speed. The speed of the attraction-repulsion movement is not uniform. It is sinusoidal as shown below. The acceleration is also sinusoidal.

5.3.1 Relationship between position, speed and acceleration

In the orbits, the position and velocity and acceleration of the movement of attraction-repulsion are perfectly related and this causes that this movement is synchronized with the one of uniform circular translation. There are two different movements that you have to distinguish very well. The orbit is eccentric and as the planet spins, simultaneously and synchronously, it is approaching and moving away from the Sun.

Instead of the apses A and P and equilibrium distance E, we use the most general notation of maximum, minimum and equilibrium position. We also use M and m as more generic than Sol and planets. This sequence will then be seen numerically in the orbits of the Moon and the planets. The sequence is as follows. If m is at a distance x from M less than x_0 it will be removed with a repulsion force given by equation (1). If m is initially farther from M than the distance x_0 will be attracted by a force also expressed by (1). In any case it will pass away or approaching the equilibrium point x_0 where the acceleration becomes equal to zero but as the body m carries a speed it will not stop. From x_0 the acceleration changes sign and the body planetts to slow down. That is, at point x_0 the velocity is maximum and decreases until reaching xmin or xmax, according to the direction the body carries. At those points it stops. In both the speed becomes equal to zero but the acceleration that in both points goes in the opposite direction to the speed, causes the body to return and the cycle repeats indefinitely.

The following table summarizes the values of acceleration and speed in a complete cycle. One way from xmin to xmax and back again.

Position	In x_{min}	To x_0	In x_0	To x_{max}	In x_{max}	To x_0	In x_0	To x_{min}
Accelerat.	(+), máx$_1$	(+), ($\downarrow$)	0	(-),($\uparrow$)	(-) máx$_2$	(-), ($\downarrow$)	0	(+), ($\uparrow$)
Speed	0	(+), ($\uparrow$)	(+), máx	(+), ($\downarrow$)	0	(-), ($\uparrow$)	(-), máx	(-), ($\downarrow$)

We will use the symbols (+) and (-) to indicate if the value is positive or negative. Positive is in the direction of increase of x, and negative on the contrary. ($\uparrow$) and ($\downarrow$) means that it is increasing or decreasing respectively in absolute value.

(0) means that at that point the value considered is zero, it is canceled. By symmetry, the maximum velocity at x_0 has the same absolute value at the going and the return. The sign is obviously different. The absolute value of the acceleration maxima, however, is not the same in xmin as in xmax. It is greater in the first case because the denominator is smaller. In the table they are represented by max1 and max2 respectively. This means that x_0 is not symmetrically located between xmin and xmax. Logically it will be closer to xmin since the braking acceleration is greater than in xmax.

5.4 Comparison of Kepler's laws with the new model

Comparing the new model with Kepler's laws results:

First law: the orbits are circular and not elliptical then the first law of Kepler is not true. Given that the eccentricity of Kepler's ellipses is small and that the foci of these ellipses are very close to the offset distance of the Sun, Kepler's first law without being conceptually true, is approximate and it seems logical that it should come out with the data available by Kepler. It is necessary to observe that the small arrows, "flechas", that leave with the new model would correspond better with vertical ellipses instead of horizontal. The Sun would be in the minor axis and off center, not in one of the foci. It should be noted that for Kepler the Earth follows an elliptical orbit with different velocities in aphelion than in perihelion. This would be true taking the Sun as a reference for the movement since $d\rho / dt$ is different in both apses. A change of speed in the translation of the Earth would imply an acceleration. This acceleration has not been observed and that implies one of two solutions: the Earth is immobile as claimed by Claudius Ptolemy or the Earth moves with uniform circular speed as it says this new model and we assume that happens in reality.

The following table compares the distance from the Sun to the center in both models. They are expressed in meters. They are very similar except for Neptune.

Position Sun		Kepler		New Modell
planet	radio	excentricidad	foco	descentrado
Mercurio	57909036500	0,20563069	11907875133	11907840900
Venus	108208104000	0,00677323	732918376	731010000
Tierra	149598261000	0,016711233	2499971396	2499971000
Marte	227939150000	0,093315	21270141782	21270150000
Ceres	414012107000	0,07582	31390397953	31340754000
Júpiter	778547186500	0,04839	37673898355	37973549500
Saturno	1433449368000	0,05648	80961220305	79876413000
Urano	2876679082500	0,044405586	127740620392	127740621500
Neptuno	4503443661500	0,00858587	38665981830	50502828500

Second law: the second law is fulfilled for any central movement in which the amount of movement is retained either the elliptical or circular orbit. As the movement of the planets is uniform circular, this law is evidence by definition of it.

Third law: Obviously Kepler refers to the cube of the semimajor axis of its ellipses and in the new model is the product of (E^2 * R) which are very approximate values, this law seems true.

$$\frac{T^2}{a^3}=k \quad \text{For Kepler}$$

$$k=\frac{w^2 R E^2}{2}*factor \quad \text{For the new model}$$

Taking into account that

$$w=\frac{2\pi}{T}$$

and that the factor is practically equal to the unit results:

$$\frac{k}{2\pi^2}=\frac{RE^2}{T^2} \quad ; \quad \frac{T^2}{RE^2}=\frac{2\pi^2}{k}$$

The new model says practically the same as Kepler's third law if you adjust the value of the constants.

In the new model it does not give exactly the same value for all planets. It may be because the planetting data is not exact. Maybe the aphelions, perihelions and periods of the different planets have to be adjusted. Or the law may not be accurate. It would be necessary to continue investigating when the values of the mass of the Sun and the gravitation constant are known. The logical thing here seems to be that Kepler's third law is fulfilled.

The following table gives the values of k that have come out with the new model for the different planets. The dimensions of k are L^3/T^2. It is measured in m^3/s^2 and its value must be multiplied by 10^9.

Planet	K *10^19
Mercury	6,321
Venus	6,635
Earth	6,634
Mars	6,571
Ceres	6,590
Jupiter	6,629
Saturn	6,704
Uranus	6,621
Neptun	6,665

6. Calculation of the orbits of the Moon and the planets

6.1 Integration of the new acceleration equation

To calculate the orbits, the following equation must be integrated twice:

$$\ddot{x}_i = k\left(\frac{1}{x_i^2} - \frac{1}{x_0^2}\right)$$

To perform these integrals, we must bear in mind that the acceleration depends on the position x, and this in turn depends on t and the functional form of that dependence is unknown, although it is known that it must be a periodic function to describe a closed orbit. That is to say:

$$x(t) = x(t+T)$$

where T is the period.

The definition of derivative is:

$$\frac{dx}{dt} = \lim_{\Delta t \to 0} \frac{x(t+\Delta t) - x(t)}{\Delta t}$$

Taking advantage of the facility offered by the spreadsheets, taking a sufficiently small Δt, the differential equation will be resolved by iteration. The traditional notation of putting a point on the derivative with respect to time and two points on the second derivative with respect to time will be used.

A spreadsheet automatically iterates if it references the previous row and is dragged down. The iteration equations are:

$$\dot{x}_i = \dot{x}_{i-1} + k\left(\frac{1}{x_i^2} - \frac{1}{x_0^2}\right)\Delta t$$

$$x = x_{i-1} + \dot{x}_i \Delta t$$

$$\left\{ \begin{array}{l} \ddot{x}_i = k\left(\frac{1}{x_i^2} - \frac{1}{x_0^2}\right) \\ \dot{x}_i = \dot{x}_{i-1} + k\left(\frac{1}{x_i^2} - \frac{1}{x_0^2}\right)\Delta t \\ x = x_{i-1} + \dot{x}_i \Delta t \end{array} \right\}$$

In the annex you can see how sequentially these formulas are introduced in the corresponding cells of the spreadsheet.

6.2 Equation of periodic movement

According to the representation that has been made and with the criterion of accepted signs, we begin to measure the time and angle in the perihelion. The origin of coordinates is taken in C, center of the orbit. The Sun is in D, to the right of C, with positive value. The perihelion is on the x-axis at + R and the aphelion on the x-axis on -R. The planet moves counter-clockwise at constant and positive angular velocity w. For each planet, T is its sidereal period.

	Phase, θ	Time, t
Perihelion	0, 2π	0, T
Aphelion	π	T/2
Any point of the orbit	θ = wt = (2π/T) t	t

$$\begin{cases} x = R\cos(wt) \\ y = R\,\text{sen}(wt) \\ v = wR = cte \\ a_t = 0 \\ a_n = w^2 R = cte \end{cases}$$

Keep in mind that the positions, speeds and accelerations are vectors. Only the modules have been placed for simplicity. The directions and senses have been described and understood to be clear.

Radio vector ρ respect to the Sun:

$$\rho(t)=\sqrt{R^2+D^2-2RD\cos(\theta t)}$$

Where, as previously stated, R is the semi-summit of aphelion and perihelion and D is the semidifference. The equation is the same for all plantes, only R, D and T vary.

6.3 Astronomical data used for the calculation

Next the astronomical data that have been used in the calculations of the orbits are put. These data have been obtained from tables that are found in appendices of astronomy books and on the Internet. The last row we added. It is the equilibrium distance E for the orbits obtained as the geometric mean of the aphelion A and perihelion P of each planet or of the apogee and perigee in the case of the Moon.

We consider that the distances and times are correct but the masses and densities are not because they were obtained with Newton's law that is not applicable to the stars. However we put them as they come but we do not use neither the masses nor the densities in our calculations.

Sun	
masa,(kg) =	1,99E+30
radio, (m) =	695508000
densidad, (kg/m³) =	1411
periodo rotación, (s) =	2356560
periodo orbital, (años) =	2,50E+08
radio orbital, (m) =	2,50E+20
velocidad orbital, (m/s) =	251000

Moon	
masa,(kg) =	7,35E+22
radio, (m) =	1737000
densidad, (kg/m³) =	3340
periodo rotación, (s) =	2360622
periodo orbital, (s) =	2360586
radio orbital, (m) =	384402000
velocidad orbital, (m/s) =	1023,1650100655
excentricidad =	0,0549
veloc. angular, (rad/s) =	2,6617057405151E-006
Apogeo, (m) =	406000000
Perigeo, (m) =	356000000
E = √AP, (m) =	380178905

Mercury	
masa,(kg) =	3,30E+23
radio, (m) =	2439700
densidad, (kg/m³) =	5430
periodo rotación, (s) =	5067000
periodo orbital, (s) =	7600428
radio orbital, (m) =	57894375900
velocidad orbital, (m/s) =	47860,6062741746
excentricidad =	0,20563069
veloc. angular, (rad/s) =	8,2668835323216E-007
Apogeo, (m) =	69816877400
Perigeo, (m) =	46001195600
E = √AP, (m) =	56671508128

Venus	
masa,(kg) =	4,87E+24
radio, (m) =	6051800
densidad, (kg/m³) =	5240
periodo rotación, (s) =	-20996815,68
periodo orbital, (s) =	19414166,4
radio orbital, (m) =	108208925000
velocidad orbital, (m/s) =	35020,6500581811
excentricidad =	0,00677323
veloc. angular, (rad/s) =	3,23639201278278E-07
Apogeo, (m) =	108939114000
Perigeo, (m) =	107477094000
E = √AP, (m) =	108205634769

Earth	
masa,(kg) =	5,97E+24
radio, (m) =	6371000
densidad, (kg/m³) =	5515
periodo rotación, (s) =	86164,1
periodo orbital, (s) =	31558149,7635456
radio orbital, (m) =	149597870691
velocidad orbital, (m/s) =	29784,7354852479
excentricidad =	0,016711233
veloc. angular, (rad/s) =	1,99909865927684E-007
Apogeo, (m) =	152098232000
Perigeo, (m) =	147098290000
E = √AP, (m) =	149577370746

Mars	
masa,(kg) =	6,42E+23
radio, (m) =	3397200
densidad, (kg/m³) =	3933,5
periodo rotación, (s) =	88642,44
periodo orbital, (s) =	59354294,4
radio orbital, (m) =	227936640000
velocidad orbital, (m/s) =	24129,1411496568
excentricidad =	0,093315
veloc. angular, (rad/s) =	1,0585898410039E-007
Apogeo, (m) =	249209300000
Perigeo, (m) =	206669000000
E = √AP, (m) =	226944567729

Ceres	
masa,(kg) =	9,43E+20
radio, (m) =	476200
densidad, (kg/m³) =	2020
periodo rotación, (s) =	32666
periodo orbital, (s) =	145324800
radio orbital, (m) =	4,138E+11
velocidad orbital, (m/s) =	17892
excentricidad =	0,07582
veloc. angular, (rad/s) =	0,00000
Apogeo, (m) =	445352861000
Perigeo, (m) =	382671353000
E = √AP, (m) =	412824153704

Jupiter	
masa,(kg) =	1,90E+27
radio, (m) =	71492000
densidad, (kg/m³) =	1336
periodo rotación, (s) =	35730
periodo orbital, (s) =	374359607
radio orbital, (m) =	7,784E+11
velocidad orbital, (m/s) =	13065
excentricidad =	0,04839
veloc. angular, (rad/s) =	1,68E-08
Apogeo, (m) =	816520736000
Perigeo, (m) =	740573637000
E = √AP, (m) =	777620557306

Saturn	
masa,(kg) =	5,69E+26
radio, (m) =	60268000
densidad, (kg/m³) =	690
periodo rotación, (s) =	38018
periodo orbital, (s) =	9,296E+08
radio orbital, (m) =	1,427E+12
velocidad orbital, (m/s) =	9643
excentricidad =	0,05648
veloc. angular, (rad/s) =	6,76E-09
Apogeo, (m) =	1513325781000
Perigeo, (m) =	1353572955000
E = √AP, (m) =	1431222152311

Uranus	
masa,(kg) =	8,69E+25
radio, (m) =	25559000
densidad, (kg/m³) =	1274
periodo rotación, (s) =	-62040
periodo orbital, (s) =	2661041808
radio orbital, (m) =	2870972217000
velocidad orbital, (m/s) =	6778,8677340372
excentricidad =	0,044405586
veloc. angular, (rad/s) =	2,3611749684993E-009
Apogeo, (m) =	3004419704000
Perigeo, (m) =	2748938461000
E = √AP, (m) =	2873841484374,5

Neptun	
masa,(kg) =	1,02E+26
radio, (m) =	24786000
densidad, (kg/m³) =	1640
periodo rotación, (s) =	57974
periodo orbital, (s) =	5200416000
radio orbital, (m) =	4498252900000
velocidad orbital, (m/s) =	5434,8260849244
excentricidad =	0,00858587
veloc. angular, (rad/s) =	1,2082082101085E-009
Apogeo, (m) =	4553946490000
Perigeo, (m) =	4452940833000
E = √AP, (m) =	4503160476445

6.4 Orbit of the Moon

Órbita Luna	
Apogeo	406000000
Perigeo	356000000
Radio	381000000
Descentrado	25000000
Periodo	2360586
w	0,000026617057405
E	380178905,2538
k	194965196086924,0
tiempo	6,5571833333
factor	0,999461332886126O
velocidad cte órbita ci	1014,1098871363
Max flecha	2941,156212
Min flecha	-2612,334673

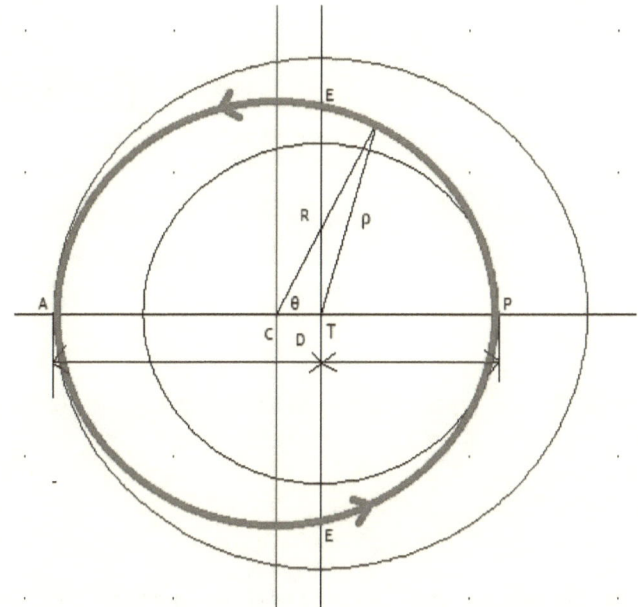

Decentral circular orbit of the Moon

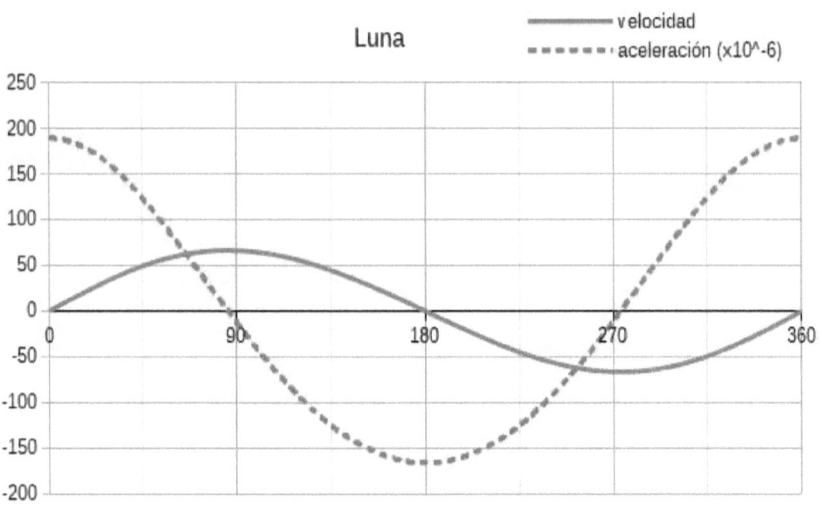

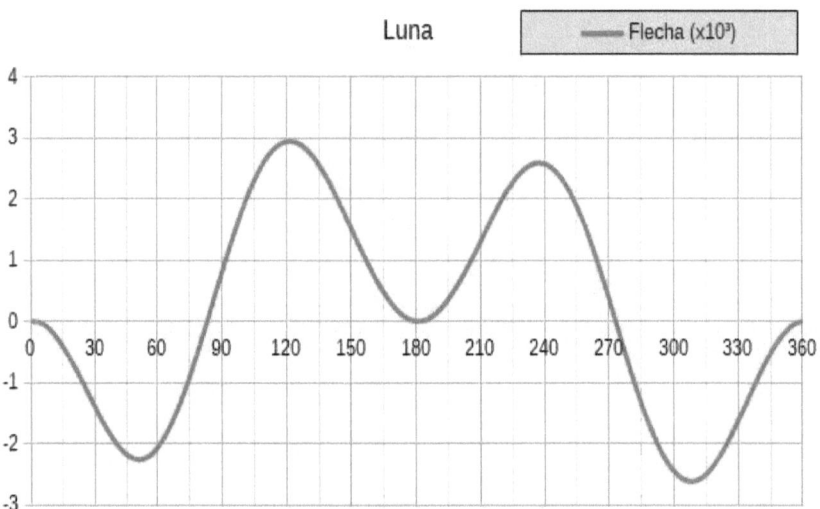

All distances are expressed in meters and all times in seconds.

θ	ρ	velocidad	aceleración	ρ circular	R	flecha
0	356000000	0,000000000000	0,000189452816	356000000	381000000	0,00
1	356000000	0,001242276850	0,000189452816	356000000	381000000	0,00
2	356000000	0,002484553699	0,000189452816	356000000	381000000	0,01
3	356000000	0,003726830547	0,000189452816	356000000	381000000	0,01
4	356000000	0,004969107394	0,000189452816	356000000	381000000	0,02

θ	ρ	velocidad	aceleración	ρ circular	R	flecha
926	356003496	1,150287274086	0,000189422602	356003494	381000002	1,89
927	356003504	1,151529352814	0,000189422537	356003502	381000002	1,89
928	356003511	1,152771431113	0,000189422471	356003509	381000002	1,89
929	356003519	1,154013508985	0,000189422406	356003517	381000002	1,89
930	356003526	1,155255586427	0,000189422340	356003524	381000002	1,89

θ	ρ	velocidad	aceleración	ρ circular	R	flecha
1855	356014021	2,303932504189	0,000189331647	356014021	381000000	0,01
1856	356014036	2,305173986509	0,000189331516	356014036	381000000	0,00
1857	356014051	2,306415467972	0,000189331386	356014051	381000000	0,00
1858	356014066	2,307656948579	0,000189331255	356014066	381000000	0,00
1859	356014082	2,308898428328	0,000189331124	356014082	381000000	-0,01

θ	ρ	velocidad	aceleración	ρ circular	R	flecha
51048	365798039	53,898314055684	0,000108145580	365800292	380997744	-2256,22
51049	365798392	53,899023186079	0,000108142764	365800645	380997744	-2256,22
51050	365798746	53,899732298012	0,000108139949	365800999	380997744	-2256,22
51051	365799099	53,900441391483	0,000108137133	365801352	380997744	-2256,22
51052	365799452	53,901150466491	0,000108134318	365801705	380997744	-2256,22

θ	ρ	velocidad	aceleración	ρ circular	R	flecha
83123	378820518	66,461771061699	0,000009691233	378820518	381000000	-0,22
83124	378820954	66,461834608893	0,000009688107	378820954	381000000	-0,10
83125	378821390	66,461898135589	0,000009684982	378821390	381000000	0,02
83126	378821826	66,461961641789	0,000009681856	378821825	381000000	0,14
83127	378822261	66,462025127491	0,000009678730	378822261	381000000	0,25

θ	ρ	velocidad	aceleración	ρ circular	R	flecha
86235	380178079	66,560589795919	0,000000005865	380177713	381000366	366,06
86236	380178515	66,560589834377	0,000000002768	380178150	381000366	366,18
86237	380178952	66,560589852526	-0,000000000329	380178586	381000366	366,30
86238	380179388	66,560589850367	-0,000000003426	380179022	381000366	366,41
86239	380179825	66,560589827899	-0,000000006523	380179459	381000367	366,53

θ	ρ	velocidad	aceleración	ρ circular	R	flecha
121531	394652635	54,756743526923	-0,000097126738	394649698	381002941	2941,16
121532	394652994	54,756106649098	-0,000097129015	394650057	381002941	2941,16
121533	394653353	54,755469756338	-0,000097131293	394650416	381002941	2941,16
121534	394653712	54,754832848643	-0,000097133571	394650775	381002941	2941,16
121535	394654071	54,754195926013	-0,000097135848	394651134	381002941	2941,16

θ	ρ	velocidad	aceleración	ρ circular	R	flecha
179998	406000000	0,002178574179	-0,000166121189	406000000	381000000	0,01
179999	406000000	0,001089287090	-0,000166121189	406000000	381000000	0,00
180000	406000000	0,000000000000	-0,000166121189	406000000	381000000	0,00
180001	406000000	-0,001089287090	-0,000166121189	406000000	381000000	0,00
180002	406000000	-0,002178574179	-0,000166121189	406000000	381000000	-0,01

θ	ρ	velocidad	aceleración	ρ circular	R	flecha
237529	394987927	-54,151516731596	-0,000099251016	394985347	381002584	2583,53
237530	394987572	-54,152167538701	-0,000099248769	394984992	381002584	2583,53
237531	394987216	-54,152818331074	-0,000099246522	394984637	381002584	2583,53
237532	394986861	-54,153469108713	-0,000099244275	394984282	381002584	2583,53
237533	394986506	-54,154119871619	-0,000099242028	394983926	381002584	2583,53

θ	ρ	velocidad	aceleración	ρ circular	R	flecha
273158	380442563	-66,556880614773	-0,000001869008	380442562	381000000	0,25
273159	380442126	-66,556892870203	-0,000001865918	380442126	381000000	0,13
273160	380441690	-66,556905105369	-0,000001862827	380441690	381000000	0,01
273161	380441253	-66,556917320270	-0,000001859737	380441253	381000000	-0,11
273162	380440817	-66,556929514905	-0,000001856646	380440817	381000000	-0,22

θ	ρ	velocidad	aceleración	ρ circular	R	flecha
273761	380179388	-66,560589827899	-0,000000003426	380179459	380999929	-70,86
273762	380178952	-66,560589850367	-0,000000000329	380179022	380999929	-70,98
273763	380178515	-66,560589852526	0,000000002768	380178586	380999929	-71,10
273764	380178079	-66,560589834377	0,000000005865	380178150	380999929	-71,21
273765	380177642	-66,560589795919	0,000000008962	380177713	380999929	-71,33

θ	ρ	velocidad	aceleración	ρ circular	R	flecha
308014	366131221	-54,555339499122	0,000105494933	366133830	380997388	-2612,33
308015	366130863	-54,554647749507	0,000105497775	366133472	380997388	-2612,33
308016	366130506	-54,553955981257	0,000105500617	366133114	380997388	-2612,33
308017	366130148	-54,553264194371	0,000105503459	366132756	380997388	-2612,33
308018	366129790	-54,552572388850	0,000105506301	366132399	380997388	-2612,33

θ	ρ	velocidad	aceleración	ρ circular	R	flecha
359998	356000000	-0,002484553699	0,000189452816	356000000	381000000	-0,01
359999	356000000	-0,001242276850	0,000189452816	356000000	381000000	0,00
360000	356000000	0,000000000000	0,000189452816	356000000	381000000	0,00
360001	356000000	0,001242276850	0,000189452816	356000000	381000000	0,00
360002	356000000	0,002484553699	0,000189452816	356000000	381000000	0,01

6.5 Orbits of the planets

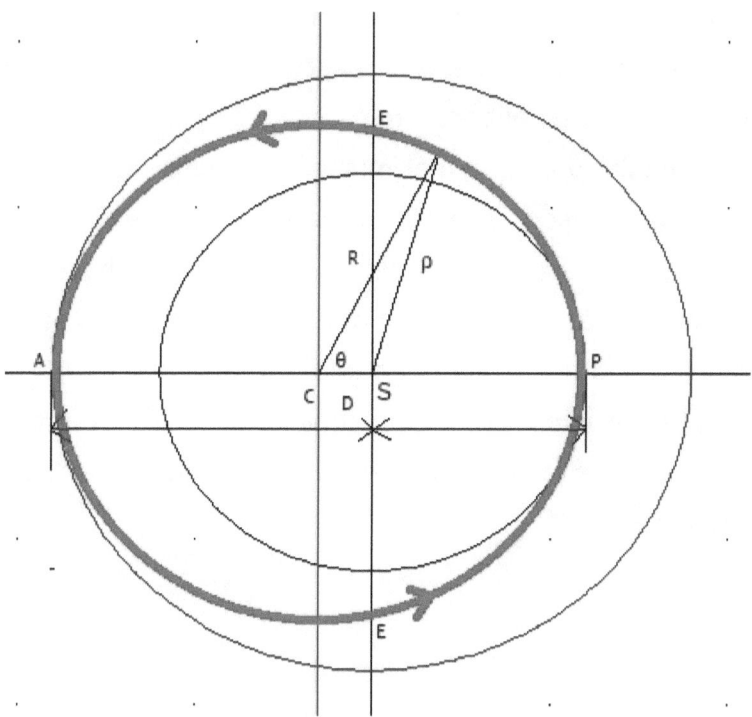

Decentral circular orbit of the planets

Next the orbit for each planet is particularized. The distances are expressed in meters and the times in seconds. We used 360000 rows in the spreadsheet then the angles θ are expressed in thousandths of sexagesimal degree. The arrows or positive or negative separation of the perfect circumference, vary with the number of iterations. Those that are put is the highest precision that our computer can give. For space reasons the 360000 rows for each planet have not been represented. The rows that have some singular point like maximums, minimums and change of sign of position, speed, acceleration and arrow have been put. The factor that multiplies the constant k has been adjusted so that the perihelion of each star coincides at the beginning and at the end. The intermediate singular points can not be ajuplanet with the intervals taken. Several rows around these singular points have been taken to facilitate interpolation.

6.5.1 Mercury Orbit

Órbita Mercurio

Afelio, A	69816877400
Perihelio, P	46001195600
Radio, R	57909036500
Descentrado del Sol, D	11907840900
Periodo, T	7600428
w	8,26688353232158E-007
E	56671508127,6175
k	63213202490729800000
incremento de tiempo	21,112300
factor	0,9946682858
velocidad cte. órbita circular	47872,7260
flecha máxima	15077140
flecha mínima	-10115734

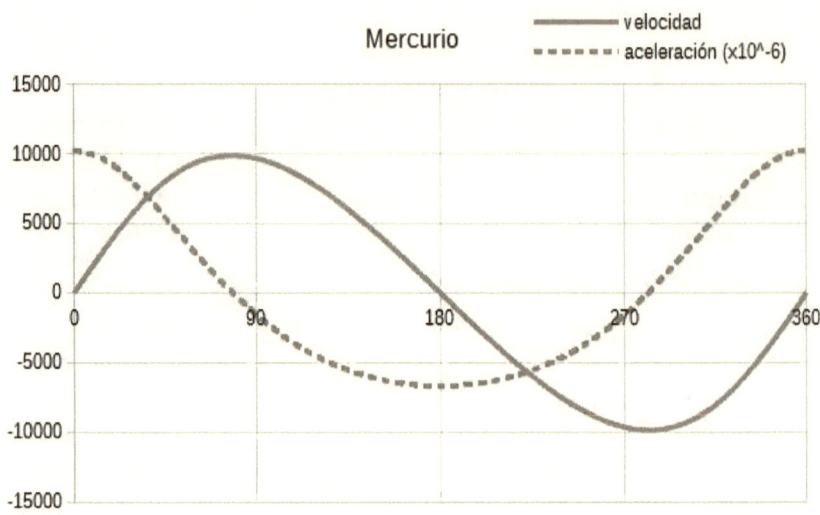

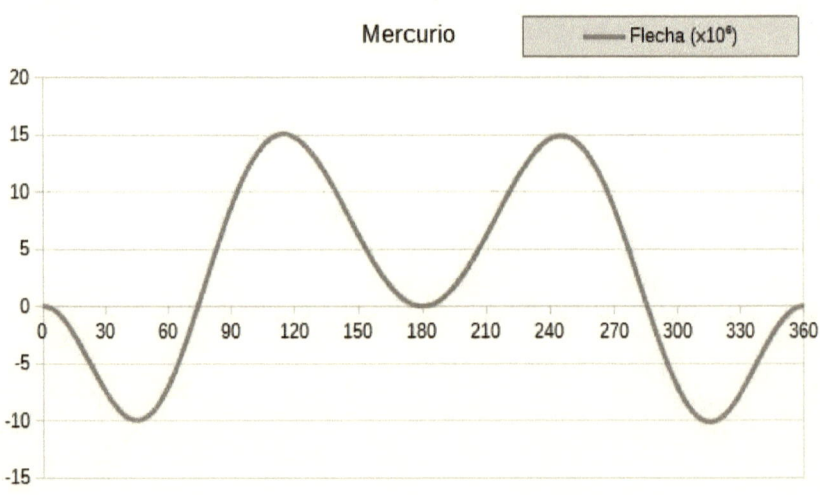

θ	ρ	velocidad	aceleración	ρ circular	R	flecha
0	46001195600	0,000000000000	0,010189952332	46001195600	57909036500	0
1	46001195605	0,215133330625	0,010189952326	46001195602	57909036502	2
2	46001195614	0,430266661126	0,010189952315	46001195609	57909036504	4
3	46001195627	0,645399991377	0,010189952297	46001195621	57909036507	7
4	46001195645	0,860533321255	0,010189952273	46001195637	57909036509	9

θ	ρ	velocidad	aceleración	ρ circular	R	flecha
91	46001214613	19,577117447151	0,010189927639	46001214507	57909036606	106
92	46001215030	19,792250256452	0,010189927097	46001214925	57909036606	106
93	46001215453	20,007383054295	0,010189926548	46001215347	57909036606	106
94	46001215880	20,222515840556	0,010189925994	46001215774	57909036606	106
95	46001216311	20,437648615110	0,010189925433	46001216205	57909036606	106

θ	ρ	velocidad	aceleración	ρ circular	R	flecha
44785	50069283843	8092,293595766260	0,006462650892	50068292724	57909090263	-9944247
44786	50069256369	8092,012665932200	0,006462889037	50069264669	57909090263	-9944247
44787	50069275438	8092,223626092960	0,006462846680	50069236039	57909090299	-9944247
44788	50069296262	8092,642828266820	0,006462846522	50069296205	57909090253	-9944243
44789	50069283625	8092,660656356780	0,006463946963	50069267920	57909090253	-9944246

θ	ρ	velocidad	aceleración	ρ circular	R	flecha
78121	56671173603	9870,671623387310	0,000000232367	56668886737	57911373302	2336802
78122	56671381996	9870,671628293110	0,000000087613	56669094568	57911373876	2337376
78123	56671590388	9870,671630142830	-0,000000057139	56669302399	57911374450	2337950
78124	56671798781	9870,671628936480	-0,000000201890	56669510230	57911375023	2338523
78125	56672007174	9870,671624674110	-0,000000346640	56669718061	57911375597	2339097

θ	ρ	velocidad	aceleración	ρ circular	R	flecha
114652	63815343241	8120,963952891590	-0,004160055370	63800484553	57924113640	15077140
114653	63815514692	8120,876124554600	-0,004160138777	63800655999	57924113640	15077140
114654	63815686140	8120,788294456710	-0,004160222181	63800827442	57924113640	15077140
114655	63815857587	8120,700462597950	-0,004160305585	63800998884	57924113640	15077140
114656	63816029032	8120,612628978350	-0,004160388986	63801170324	57924113640	15077140

θ	ρ	velocidad	aceleración	ρ circular	R	flecha
180185	69816825912	-26,223322598668	-0,006713973360	69816825914	57909036498	-2
180186	69816825356	-26,365070018442	-0,006713973154	69816825356	57909036499	-1
180187	69816824796	-26,506817433851	-0,006713972946	69816824795	57909036501	1
180188	69816824233	-26,648564844871	-0,006713972737	69816824231	57909036502	2
180189	69816823668	-26,790312251478	-0,006713972526	69816823664	57909036504	4

θ	ρ	velocidad	aceleración	ρ circular	R	flecha
245250	63831965288	-8112,348411363050	-0,004168138458	63817277502	57923939763	14903263
245251	63831794016	-8112,436410352620	-0,004168055203	63817106235	57923939763	14903263
245252	63831622742	-8112,524407584490	-0,004167971946	63816934965	57923939763	14903263
245253	63831451466	-8112,612403058610	-0,004167888688	63816763694	57923939763	14903263
245254	63831280189	-8112,700396774950	-0,004167805428	63816592421	57923939763	14903263

θ	ρ	velocidad	aceleración	ρ circular	R	flecha
281875	56671798781	-9870,671624674120	-0,000000201890	56669718061	57911162654	2126154
281876	56671590388	-9870,671628936490	-0,000000057139	56669510230	57911162080	2125580
281877	56671381996	-9870,671630142830	0,000000087613	56669302399	57911161507	2125007
281878	56671173603	-9870,671628293120	0,000000232367	56669094568	57911160933	2124433
281879	56670965211	-9870,671623387310	0,000000377123	56668886737	57911160359	2123859

θ	ρ	velocidad	aceleración	ρ circular	R	flecha
285589	55898384589	-9849,332232227110	0,000548214964	55898383470	57909037644	1144
285590	55898176647	-9849,320658148320	0,000548365481	55898176084	57909037076	576
285591	55897968706	-9849,309080891770	0,000548515999	55897968697	57909036509	9
285592	55897760765	-9849,297500457450	0,000548666519	55897761312	57909035941	-559
285593	55897552823	-9849,285916845310	0,000548817040	55897553926	57909035373	-1127

θ	ρ	velocidad	aceleración	ρ circular	R	flecha
315142	50166094765	-8013,906262299730	0,005435715236	50176067736	57898920766	-10115734
315143	50165925575	-8013,791501848960	0,005435884663	50175898551	57898920766	-10115734
315144	50165756388	-8013,676737821190	0,005436054089	50175729368	57898920766	-10115734
315145	50165587203	-8013,561970216450	0,005436223515	50175560187	57898920766	-10115734
315146	50165418021	-8013,447199034740	0,005436392940	50175391009	57898920766	-10115734

θ	ρ	velocidad	aceleración	ρ circular	R	flecha
359998	46001195605	-0,430266661119	0,010189952326	46001195609	57909036495	-5
359999	46001195600	-0,215133330619	0,010189952332	46001195602	57909036498	-2
360000	46001195600	0,000000000006	0,010189952332	46001195600	57909036500	0
360001	46001195605	0,215133330631	0,010189952326	46001195602	57909036502	2
360002	46001195614	0,430266661132	0,010189952315	46001195609	57909036504	4

6.5.2 Orbit of Venus

Órbita Venus	
Afelio, A	108939114000
Perihelio, P	107477094000
Radio, R	108208104000
Descentrado del Sol, D	731010000
Periodo, T	19414166,4
w	3,23639201278278E-007
E	108205634768,5030
k	66351280674631100000
incremento de tiempo	53,928240
factor	0,9999942952
velocidad cte. órbita circular	35020,3844
flecha máxima	6671
flecha mínima	-6661

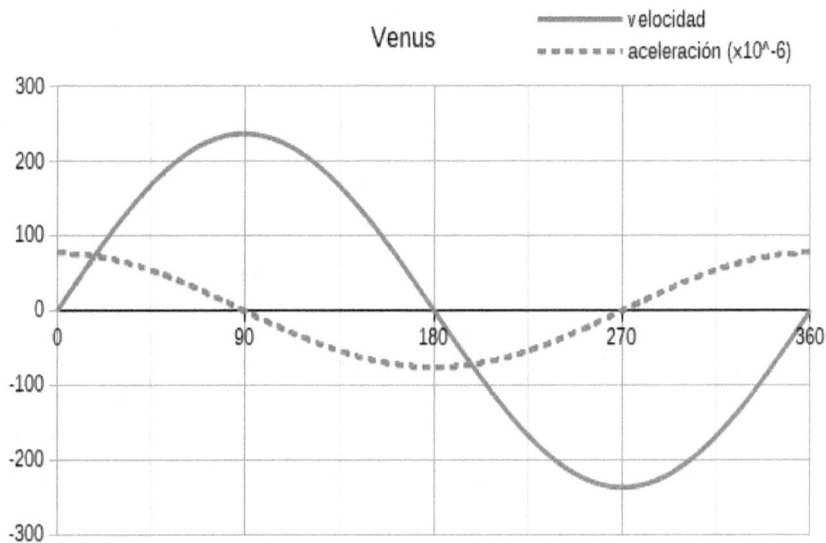

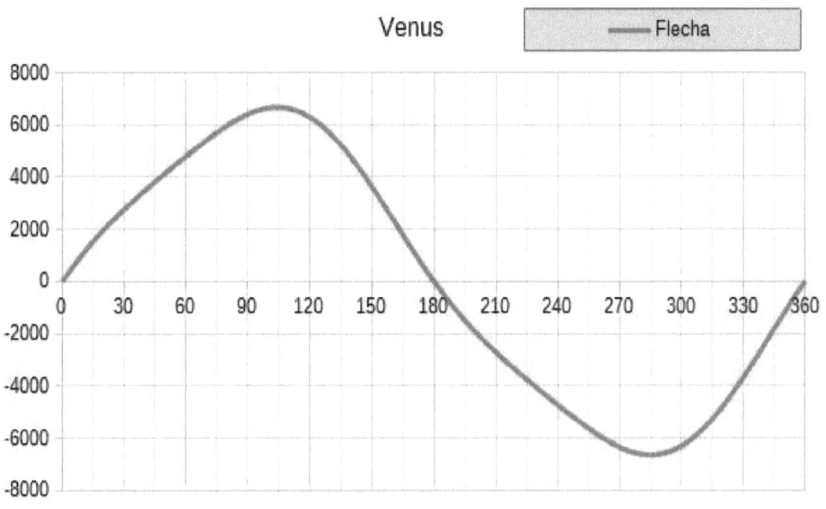

θ	ρ	velocidad	aceleración	ρ circular	R	flecha
0	107477094000	0,000000000000	0,000077088031	107477094000	108208104000	0
1	107477094000	0,004157221847	0,000077088031	107477094000	108208104000	0
2	107477094001	0,008314443692	0,000077088031	107477094000	108208104000	0
3	107477094001	0,012471665534	0,000077088031	107477094001	108208104000	0
4	107477094002	0,016628887373	0,000077088031	107477094002	108208104000	0

θ	ρ	velocidad	aceleración	ρ circular	R	flecha
89610	108205603769	236,584167064906	0,000000003247	108205597385	108208110384	6384
89611	108205616528	236,584167240010	0,000000001911	108205610144	108208110384	6384
89612	108205629286	236,584167343045	0,000000000574	108205622903	108208110384	6384
89613	108205642045	236,584167374011	-0,000000000762	108205635661	108208110384	6384
89614	108205654804	236,584167332908	-0,000000002099	108205648420	108208110384	6384
89615	108205667562	236,584167219736	-0,000000003435	108205661178	108208110384	6384

θ	ρ	velocidad	aceleración	ρ circular	R	flecha
104381	108391983639	228,782052694086	-0,000019468642	108391976968	108208110671	6671
104382	108391995976	228,781002784466	-0,000019469928	108391989306	108208110671	6671
104383	108392008314	228,779952805513	-0,000019471214	108392001644	108208110671	6671
104384	108392020652	228,778902757228	-0,000019472499	108392013981	108208110671	6671
104385	108392032989	228,777852639610	-0,000019473785	108392026319	108208110671	6671

θ	ρ	velocidad	aceleración	ρ circular	R	flecha
270385	108205654804	-236,584167219739	-0,000000002099	108205661178	108208097625	-6375
270386	108205642045	-236,584167332910	-0,000000000762	108205648420	108208097625	-6375
270387	108205629286	-236,584167374013	0,000000000574	108205635661	108208097625	-6375
270388	108205616528	-236,584167343047	0,000000001911	108205622903	108208097625	-6375
270389	108205603769	-236,584167240012	0,000000003247	108205610144	108208097625	-6375

θ	ρ	velocidad	aceleración	ρ circular	R	flecha
285111	108019835878	-228,801647041339	0,000019511583	108019842538	108208097339	-6661
285112	108019823539	-228,800594816028	0,000019512882	108019830199	108208097339	-6661
285113	108019811200	-228,799542520659	0,000019514181	108019817861	108208097339	-6661
285114	108019798861	-228,798490155232	0,000019515480	108019805522	108208097339	-6661
285115	108019786523	-228,797437719748	0,000019516779	108019793183	108208097339	-6661
285116	108019774184	-228,796385214207	0,000019518078	108019780845	108208097339	-6661
285117	108019761846	-228,795332638609	0,000019519377	108019768506	108208097339	-6661

θ	ρ	velocidad	aceleración	ρ circular	R	flecha
359998	107477094000	-0,008314443692	0,000077088031	107477094000	108208104000	0
359999	107477094000	-0,004157221846	0,000077088031	107477094000	108208104000	0
360000	107477094000	0,000000000000	0,000077088031	107477094000	108208104000	0
360001	107477094000	0,004157221847	0,000077088031	107477094000	108208104000	0
360002	107477094001	0,008314443692	0,000077088031	107477094000	108208104000	0

6.5.3 Earth orbit

Órbita Tierra	
Afelio, A	152098232000
Perihelio, P	147098290000
Radio, R	149598261000
Descentrado del Sol, D	2499971000
Periodo, T	31558149,763546
w	1,99098659276835E-007
E	149577370745,7890
k	66336086829341800000
incremento de tiempo	87,661527121
factor	0,9999650898
velocidad cte. órbita circular	29784,8132
flecha máxima	35545
flecha mínima	-34998

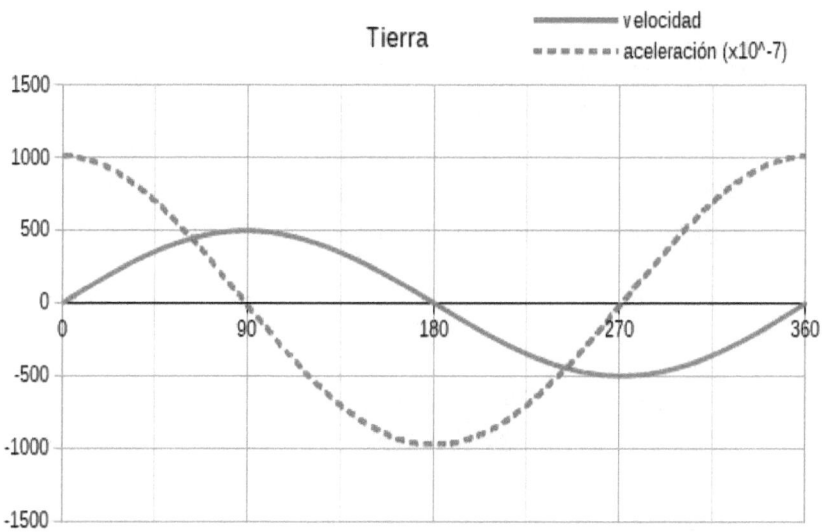

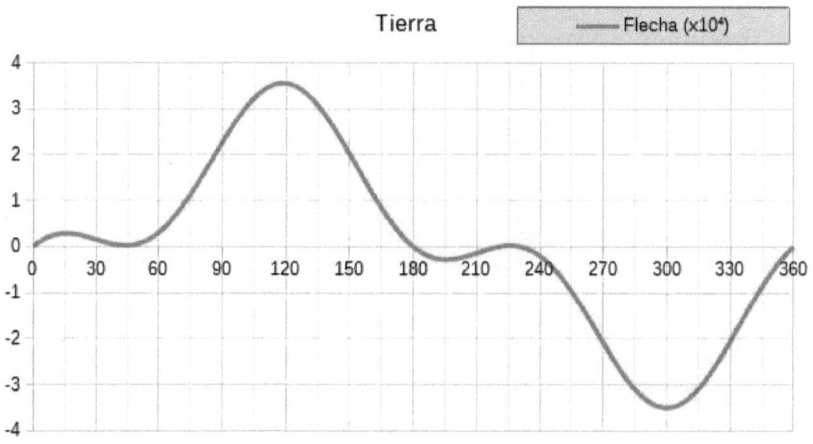

θ	ρ	velocidad	aceleración	ρ circular	R	flecha
0	147098290000	0,000000000000	0,000100780243	147098290000	149598261000	0
1	147098290001	0,008834550007	0,000100780243	147098290000	149598261000	0
2	147098290002	0,017669100011	0,000100780243	147098290002	149598261001	1
3	147098290005	0,026503650009	0,000100780243	147098290003	149598261001	1
4	147098290008	0,035338199999	0,000100780243	147098290006	149598261002	2

θ	ρ	velocidad	aceleración	ρ circular	R	flecha
15738	147193573542	137,208021930629	0,000096812413	147193570663	149598263879	2879
15739	147193585571	137,216508654604	0,000096811913	147193582692	149598263879	2879
15740	147193597600	137,224995334712	0,000096811412	147193594721	149598263879	2879
15741	147193609630	137,233481970950	0,000096810912	147193606751	149598263879	2879
15742	147193621661	137,241968563317	0,000096810411	147193618782	149598263879	2879

θ	ρ	velocidad	aceleración	ρ circular	R	flecha
44095	147813051527	350,533001759573	0,000071202728	147813051287	149598261240	240
44096	147813082256	350,539243499480	0,000071201466	147813082016	149598261240	240
44097	147813112985	350,545485128726	0,000071200204	147813112745	149598261240	240
44098	147813143715	350,551726647309	0,000071198941	147813143475	149598261240	240
44099	147813174446	350,557968055227	0,000071197679	147813174205	149598261240	240
44100	147813205177	350,564209352477	0,000071196416	147813204936	149598261240	240

θ	ρ	velocidad	aceleración	ρ circular	R	flecha
89040	149577284924	497,749562740467	0,000000003402	149577262866	149598283061	22061
89041	149577328557	497,749563038725	0,000000001673	149577306498	149598283062	22062
89042	149577372191	497,749563185343	-0,000000000057	149577350131	149598283063	22063
89043	149577415824	497,749563180322	-0,000000001787	149577393764	149598283063	22063
89044	149577459458	497,749563023662	-0,000000003517	149577437397	149598283064	22064

θ	ρ	velocidad	aceleración	ρ circular	R	flecha
118481	150806461170	433,999058314938	-0,000048132514	150806425629	149598296545	35545
118482	150806499215	433,994838945221	-0,000048133986	150806463674	149598296545	35545
118483	150806537259	433,990619446496	-0,000048135458	150806501718	149598296545	35545
118484	150806575303	433,986399818761	-0,000048136929	150806539762	149598296545	35545
118485	150806613346	433,982180062020	-0,000048138401	150806577806	149598296545	35545
118486	150806651389	433,977960176273	-0,000048139873	150806615849	149598296545	35545

θ	ρ	velocidad	aceleración	ρ circular	R	flecha
195625	152007334485	-131,932768613079	-0,000094036851	152007337263	149598258223	-2777
195626	152007322919	-131,941012027043	-0,000094036414	152007325697	149598258223	-2777
195627	152007311352	-131,949255402709	-0,000094035977	152007314130	149598258223	-2777
195628	152007299785	-131,957498740074	-0,000094035540	152007302562	149598258223	-2777
195629	152007288216	-131,965742039135	-0,000094035103	152007290994	149598258223	-2777

θ	ρ	velocidad	aceleración	ρ circular	R	flecha
220528	151507169926	-319,362296931680	-0,000075050270	151507169927	149598261000	0
220529	151507141930	-319,368875953003	-0,000075049202	151507141930	149598261000	0
220530	151507113933	-319,375454880701	-0,000075048134	151507113933	149598261000	0
220531	151507085936	-319,382033714772	-0,000075047066	151507085936	149598261000	0
220532	151507057938	-319,388612455214	-0,000075045998	151507057937	149598261000	0

θ	ρ	velocidad	aceleración	ρ circular	R	flecha
270955	149577459458	-497,749562715360	-0,000000003517	149577481029	149598239425	-21575
270956	149577415824	-497,749563023659	-0,000000001787	149577437397	149598239424	-21576
270957	149577372191	-497,749563180319	-0,000000000057	149577393764	149598239424	-21576
270958	149577328557	-497,749563185340	0,000000001673	149577350131	149598239423	-21577
270959	149577284924	-497,749563038722	0,000000003402	149577306498	149598239422	-21578
270960	149577241290	-497,749562740465	0,000000005132	149577262866	149598239421	-21579

θ	ρ	velocidad	aceleración	ρ circular	R	flecha
300143	148358593602	-434,032462335334	0,000048914754	148358628596	149598226002	-34998
300144	148358555554	-434,028174393282	0,000048916300	148358590548	149598226002	-34998
300145	148358517507	-434,023886315718	0,000048917846	148358552501	149598226002	-34998
300146	148358479460	-434,019598102643	0,000048919392	148358514454	149598226002	-34998
300147	148358441414	-434,015309754059	0,000048920938	148358476408	149598226002	-34998
300148	148358403368	-434,011021269966	0,000048922483	148358438362	149598226002	-34998

θ	ρ	velocidad	aceleración	ρ circular	R	flecha
359998	147098290001	-0,017669100010	0,000100780243	147098290002	149598260999	-1
359999	147098290000	-0,008834550006	0,000100780243	147098290000	149598261000	0
360000	147098290000	0,000000000001	0,000100780243	147098290000	149598261000	0
360001	147098290001	0,008834550008	0,000100780243	147098290000	149598261000	0
360002	147098290002	0,017669100012	0,000100780243	147098290002	149598261001	1

6.5.4 Mars Orbit

Órbita Marte	
Afelio, A	249209300000
Perihelio, P	206669000000
Radio, R	227939150000
Descentrado del Sol, D	21270150000
Periodo, T	59354294,4
w	1,05858984100392E-007
E	226944567728,9940
k	65706773999652300000
incremento de tiempo	164,873040
factor	0,9989096055
velocidad cte. órbita circular	24129,4069
flecha máxima	5040512
flecha mínima	-4233883

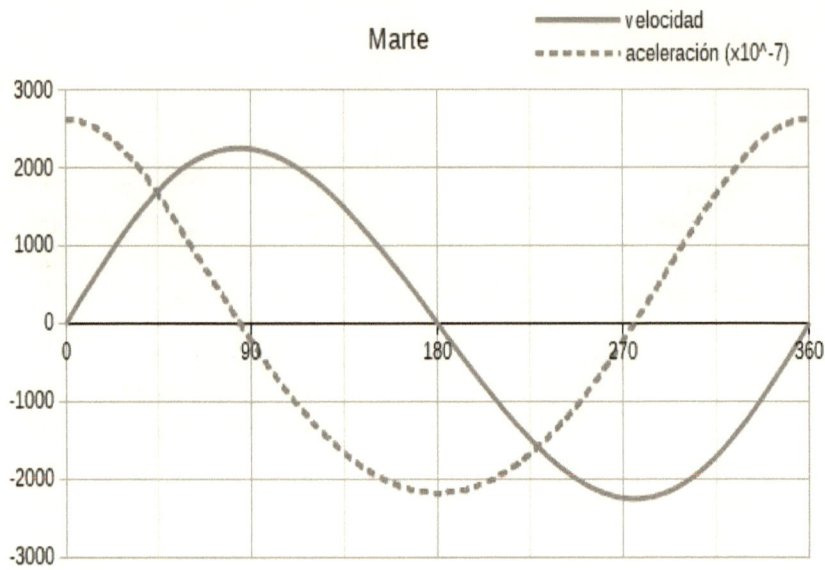

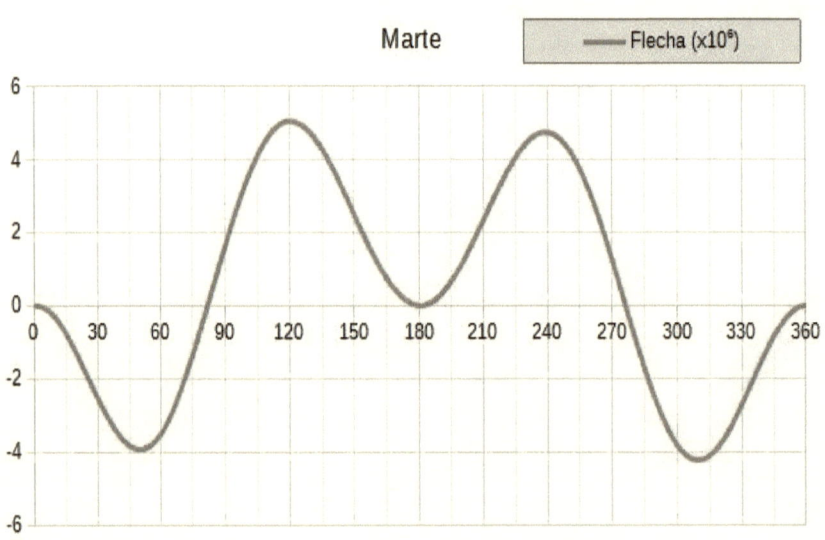

θ	ρ	velocidad	aceleración	ρ circular	R	flecha
0	206669000000	0,000000000000	0,000262600635	206669000000	227939150000	0
1	206669000007	0,043295765007	0,000262600635	206669000004	227939150004	4
2	206669000021	0,086591529997	0,000262600635	206669000014	227939150007	7
3	206669000043	0,129887294951	0,000262600634	206669000032	227939150011	11
4	206669000071	0,173183059853	0,000262600634	206669000057	227939150014	14

θ	ρ	velocidad	aceleración	ρ circular	R	flecha
456	206669743778	19,742591959885	0,000262589562	206669742960	227939150817	817
457	206669747040	19,785885899297	0,000262589514	206669746222	227939150817	817
458	206669750309	19,829179830702	0,000262589465	206669749492	227939150817	817
459	206669753585	19,872473754083	0,000262589416	206669752768	227939150817	817
460	206669756869	19,915767669421	0,000262589367	206669756052	227939150817	817

θ	ρ	velocidad	aceleración	ρ circular	R	flecha
914	206671984833	39,570099587074	0,000262556200	206671984825	227939150008	8
915	206671991365	39,613388025954	0,000262556103	206671991360	227939150005	5
916	206671997903	39,656676448803	0,000262556006	206671997902	227939150001	1
917	206672004448	39,699964855605	0,000262555908	206672004451	227939149997	-3
918	206672011001	39,743253246342	0,000262555811	206672011007	227939149994	-6
919	206672017561	39,786541620996	0,000262555713	206672017570	227939149990	-10

θ	ρ	velocidad	aceleración	ρ circular	R	flecha
49948	214866000088	1828,361310656660	0,000147464234	214869919408	227935219379	-3930621
49949	214866301540	1828,385623533190	0,000147460240	214870220860	227935219379	-3930621
49950	214866602995	1828,409935751310	0,000147456247	214870522315	227935219379	-3930621
49951	214866904455	1828,434247310990	0,000147452253	214870823774	227935219379	-3930621
49952	214867205918	1828,458558212250	0,000147448260	214871125237	227935219379	-3930621

θ	ρ	velocidad	aceleración	ρ circular	R	flecha
81945	225942365408	2250,347694772000	0,000011342804	225942365790	227939149616	-384
81946	225942736430	2250,349564894520	0,000011338577	225942736610	227939149819	-181
81947	225943107452	2250,351434320100	0,000011334349	225943107430	227939150022	22
81948	225943478475	2250,353303048750	0,000011330122	225943478251	227939150224	224
81949	225943849497	2250,355171080460	0,000011325895	225943849072	227939150427	427

θ	ρ	velocidad	aceleración	ρ circular	R	flecha
84642	226943758012	2252,867411367280	0,000000009104	226943211856	227939698550	548550
84643	226944129449	2252,867412868230	0,000000004928	226943583090	227939698753	548753
84644	226944500886	2252,867413680660	0,000000000752	226943954324	227939698957	548957
84645	226944872323	2252,867413804560	-0,000000003425	226944325558	227939699161	549161
84646	226945243760	2252,867413239950	-0,000000007601	226944696793	227939699365	549365

θ	ρ	velocidad	aceleración	ρ circular	R	flecha
120354	239397519507	1850,035445528450	-0,000129273107	239392493831	227944190512	5040512
120355	239397824524	1850,014131878260	-0,000129276029	239392798848	227944190512	5040512
120356	239398129538	1849,992817746390	-0,000129278950	239393103862	227944190512	5040512
120357	239398434549	1849,971503132860	-0,000129281872	239393408872	227944190512	5040512
120358	239398739556	1849,950188037660	-0,000129284793	239393713878	227944190512	5040512

θ	ρ	velocidad	aceleración	ρ circular	R	flecha
179998	249209299994	0,071810261151	-0,000217774419	249209299988	227939150006	6
179999	249209300000	0,035905130579	-0,000217774420	249209299997	227939150003	3
180000	249209300000	-0,000000000002	-0,000217774420	249209300000	227939150000	0
180001	249209299994	-0,035905130582	-0,000217774419	249209299997	227939149997	-3
180002	249209299982	-0,071810261154	-0,000217774419	249209299988	227939149994	-6

θ	ρ	velocidad	aceleración	ρ circular	R	flecha
239184	239537758664	-1840,137365790660	-0,000130615158	239533037044	227943885410	4735410
239185	239537455272	-1840,158900708870	-0,000130612257	239532733652	227943885410	4735410
239186	239537151876	-1840,180435148820	-0,000130609357	239532430256	227943885410	4735410
239187	239536848476	-1840,201969110490	-0,000130606456	239532126857	227943885410	4735410
239188	239536545073	-1840,223502593870	-0,000130603555	239531823454	227943885410	4735410

θ	ρ	velocidad	aceleración	ρ circular	R	flecha
275353	226945243760	-2252,867411986820	-0,000000007601	226945068027	227939326504	176504
275354	226944872323	-2252,867413239950	-0,000000003425	226944696793	227939326300	176300
275355	226944500886	-2252,867413804560	0,000000000752	226944325558	227939326096	176096
275356	226944129449	-2252,867413680660	0,000000004928	226943954324	227939325892	175892
275357	226943758012	-2252,867412868230	0,000000009104	226943583090	227939325689	175689

θ	ρ	velocidad	aceleración	ρ circular	R	flecha
276217	226624334245	-2252,611447794130	0,000003608008	226624333833	227939150414	414
276218	226623962850	-2252,610852930890	0,000003612201	226623962641	227939150210	210
276219	226623591456	-2252,610257376280	0,000003616395	226623591449	227939150006	6
276220	226623220061	-2252,609661130300	0,000003620588	226623220258	227939149803	-197
276221	226622848667	-2252,609064192960	0,000003624781	226622849066	227939149599	-401

θ	ρ	velocidad	aceleración	ρ circular	R	flecha
309585	215006908958	-1839,643680354630	0,000145599365	215011130518	227934916117	-4233883
309586	215006605654	-1839,619674944710	0,000145603375	215010827215	227934916117	-4233883
309587	215006302354	-1839,595668873620	0,000145607385	215010523915	227934916117	-4233883
309588	215005999058	-1839,571662141380	0,000145611395	215010220620	227934916117	-4233883
309589	215005695767	-1839,547654747970	0,000145615405	215009917328	227934916117	-4233883

θ	ρ	velocidad	aceleración	ρ circular	R	flecha
359998	206669000007	-0,086591529993	0,000262600635	206669000014	227939149993	-7
359999	206669000000	-0,043295765004	0,000262600635	206669000004	227939149996	-4
360000	206669000000	0,000000000003	0,000262600635	206669000000	227939150000	0
360001	206669000007	0,043295765010	0,000262600635	206669000004	227939150004	4
360002	206669000021	0,086591530000	0,000262600635	206669000014	227939150007	7

6.5.5 Orbit of Ceres

Órbita Ceres	
Afelio, A	445352861000
Perihelio, P	382671353000
Radio, R	414012107000
Descentrado del Sol, D	31340754000
Periodo, T	145324800
w	4,3235465021659E-008
E	412824153703,8390
k	65899473918679300000
incremento de tiempo	403,680000
factor	0,9992828511
velocidad cte. órbita circular	17900,0060
flecha máxima	4880129
flecha mínima	-4246773

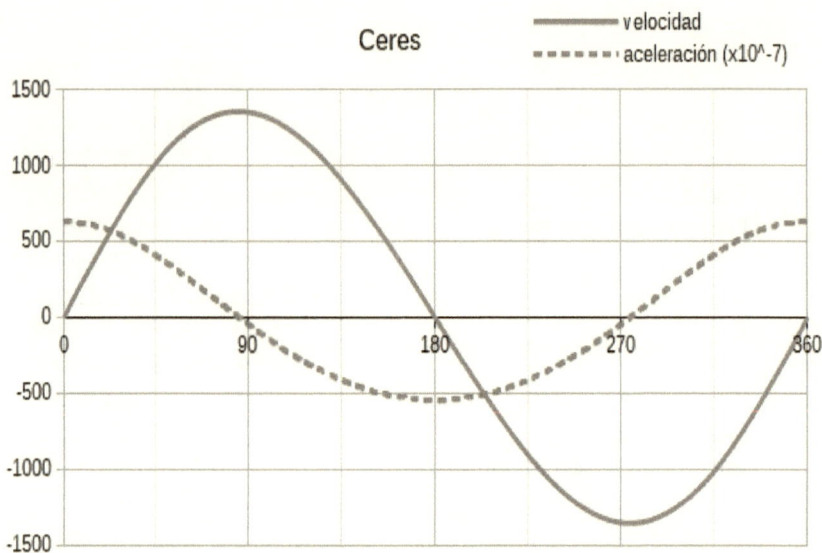
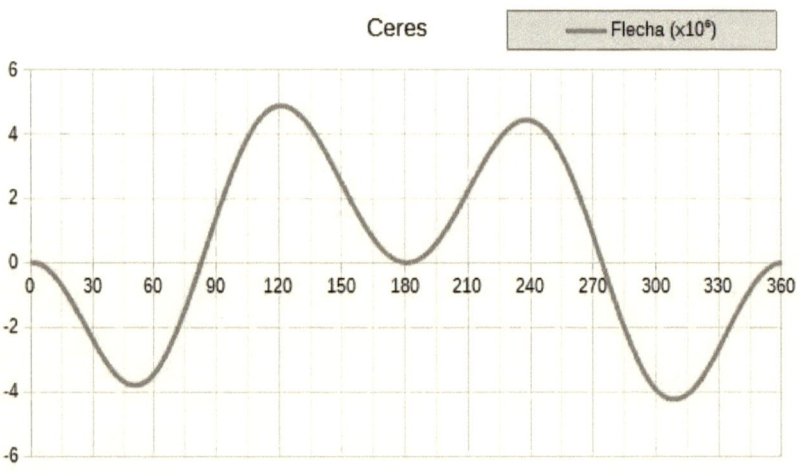

θ	ρ	velocidad	aceleración	ρ circular	R	flecha
0	382671353000	0,000000000000	0,000063338129	382671353000	414012107000	0
1	382671353010	0,025568336102	0,000063338129	382671353005	414012107005	5
2	382671353031	0,051136672194	0,000063338129	382671353021	414012107010	10
3	382671353062	0,076705008267	0,000063338129	382671353046	414012107015	15
4	382671353103	0,102273344310	0,000063338129	382671353083	414012107021	21

θ	ρ	velocidad	aceleración	ρ circular	R	flecha
695	382673849302	17,769445305450	0,000063332258	382673847504	414012108798	1798
696	382673856485	17,795011271466	0,000063332241	382673854687	414012108798	1798
697	382673863679	17,820577230662	0,000063332224	382673861881	414012108798	1798
698	382673870883	17,846143183028	0,000063332208	382673869085	414012108798	1798
699	382673878097	17,871709128554	0,000063332191	382673876300	414012108798	1798

θ	ρ	velocidad	aceleración	ρ circular	R	flecha
1392	382681359292	35,586718803289	0,000063314596	382681359281	414012107011	11
1393	382681373668	35,612277639307	0,000063314562	382681373662	414012107005	5
1394	382681388054	35,637836461677	0,000063314528	382681388054	414012107000	0
1395	382681402451	35,663395270389	0,000063314494	382681402455	414012106995	-5
1396	382681416857	35,688954065433	0,000063314460	382681416868	414012106990	-10

θ	ρ	velocidad	aceleración	ρ circular	R	flecha
50678	394893348277	1098,982256449240	0,000035912977	394897141324	414008306784	-3800216
50679	394893791920	1098,996753799670	0,000035912027	394897584966	414008306784	-3800216
50680	394894235569	1099,011250766800	0,000035911078	394898028615	414008306784	-3800216
50681	394894679224	1099,025747350620	0,000035910128	394898472270	414008306784	-3800216
50682	394895122884	1099,040243551130	0,000035909179	394898915930	414008306784	-3800216

θ	ρ	velocidad	aceleración	ρ circular	R	flecha
82815	411269372957	1353,840485887130	0,000002929171	411269373358	414012106598	-402
82816	411269919476	1353,841668334720	0,000002928135	411269919681	414012106794	-206
82817	411270465995	1353,842850364310	0,000002927100	411270466005	414012106990	-10
82818	411271012515	1353,844031975910	0,000002926064	411271012330	414012107186	186
82819	411271559035	1353,845213169510	0,000002925029	411271558655	414012107382	382

θ	ρ	velocidad	aceleración	ρ circular	R	flecha
85656	412823315166	1355,518825291940	0,000000001571	412822758253	414012665516	558516
85657	412823862362	1355,518825926070	0,000000000546	412823305253	414012665713	558713
85658	412824409558	1355,518826146390	-0,000000000479	412823852252	414012665910	558910
85659	412824956754	1355,518825952900	-0,000000001504	412824399251	414012666107	559107
85660	412825503950	1355,518825345610	-0,000000002529	412824946251	414012666304	559304

θ	ρ	velocidad	aceleración	ρ circular	R	flecha
121133	431056573040	1114,023262682090	-0,000032019060	431051702371	414016987129	4880129
121134	431057022744	1114,010337228060	-0,000032019800	431052152075	414016987129	4880129
121135	431057472443	1113,997411475300	-0,000032020540	431052601773	414016987129	4880129
121136	431057922136	1113,984485423820	-0,000032021280	431053051466	414016987129	4880129
121137	431058371824	1113,971559073620	-0,000032022020	431053501154	414016987129	4880129

θ	ρ	velocidad	aceleración	ρ circular	R	flecha
179998	445352860991	0,043939404415	-0,000054423559	445352860982	414012107009	9
179999	445352861000	0,021969702130	-0,000054423559	445352860996	414012107005	5
180000	445352861000	-0,000000000160	-0,000054423559	445352861000	414012107000	0
180001	445352860991	-0,021969702450	-0,000054423559	445352860996	414012106996	-4
180002	445352860974	-0,043939404734	-0,000054423559	445352860982	414012106991	-9

θ	ρ	velocidad	aceleración	ρ circular	R	flecha
238167	431369636200	-1104,902612637370	-0,000032533658	431365213270	414016538381	4431381
238168	431369190168	-1104,915745824360	-0,000032532925	431364767238	414016538381	4431381
238169	431368744130	-1104,928878715710	-0,000032532193	431364321200	414016538381	4431381
238170	431368298087	-1104,942011311410	-0,000032531461	431363875157	414016538381	4431381
238171	431367852039	-1104,955143611460	-0,000032530728	431363429109	414016538381	4431381

θ	ρ	velocidad	aceleración	ρ circular	R	flecha
274340	412824956754	-1355,518825345630	-0,000000001504	412824946251	414012117533	10533
274341	412824409558	-1355,518825952920	-0,000000000479	412824399251	414012117336	10336
274342	412823862362	-1355,518826146410	0,000000000546	412823852252	414012117139	10139
274343	412823315166	-1355,518825926090	0,000000001571	412823305253	414012116942	9942
274344	412822767971	-1355,518825291960	0,000000002596	412822758253	414012116745	9745

θ	ρ	velocidad	aceleración	ρ circular	R	flecha
274391	412797049770	-1355,518328682650	0,000000050780	412797049287	414012107483	483
274392	412796502574	-1355,518308183820	0,000000051805	412796502288	414012107286	286
274393	412795955378	-1355,518287271110	0,000000052830	412795955289	414012107089	89
274394	412795408183	-1355,518265944500	0,000000053856	412795408290	414012106892	-108
274395	412794860987	-1355,518244204010	0,000000054881	412794861291	414012106695	-305

θ	ρ	velocidad	aceleración	ρ circular	R	flecha
308616	395207559851	-1109,121777062070	0,000035241275	395211798464	414007860227	-4246773
308617	395207112127	-1109,107550864250	0,000035242231	395211350740	414007860227	-4246773
308618	395206664408	-1109,093324280530	0,000035243187	395210903022	414007860227	-4246773
308619	395206216695	-1109,079097310890	0,000035244143	395210455309	414007860227	-4246773
308620	395205768988	-1109,064869955360	0,000035245099	395210007602	414007860227	-4246773

θ	ρ	velocidad	aceleración	ρ circular	R	flecha
359998	382671353010	-0,051136671836	0,000063338129	382671353021	414012106990	-10
359999	382671353000	-0,025568335744	0,000063338129	382671353005	414012106995	-5
360000	382671353000	0,000000000358	0,000063338129	382671353000	414012107000	0
360001	382671353010	0,025568336460	0,000063338129	382671353005	414012107005	5
360002	382671353031	0,051136672553	0,000063338129	382671353021	414012107010	10

6.5.6 Jupiter's orbit

Órbita Júpiter	
Afelio, A	816520736000
Perihelio, P	740573637000
Radio, R	778547186500
Descentrado del Sol, D	37973549500
Periodo, T	374359607
w	1,6783822799503E-008
E	777620557306,3490
k	66289226602267000000
incremento de tiempo	1039,887797
factor	0,9997024824
velocidad cte. órbita circular	13066,9980
flecha máxima	2551879
flecha mínima	-2346740

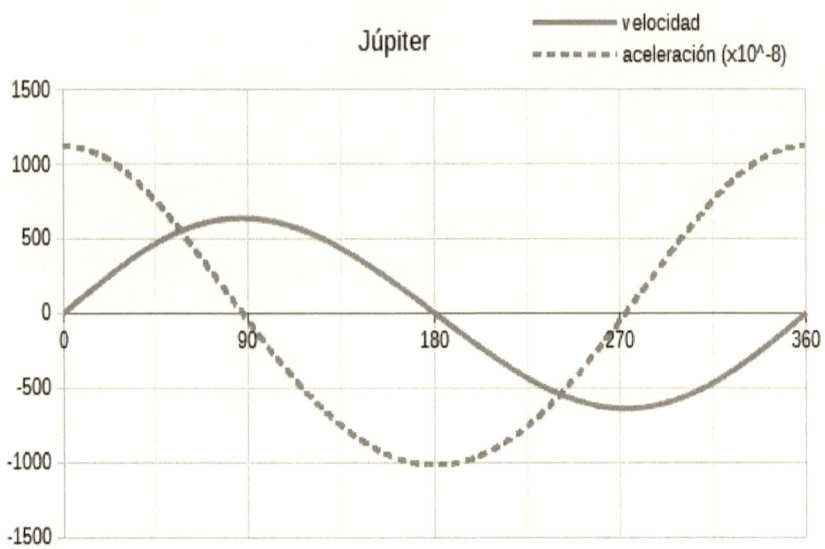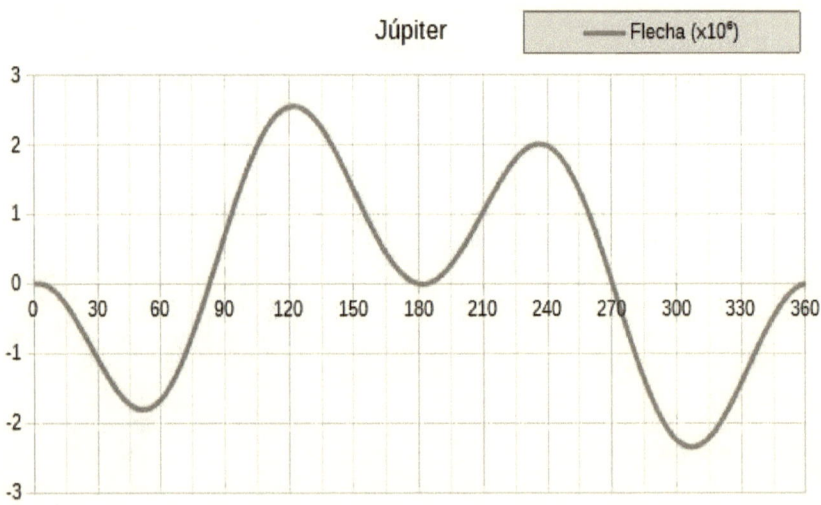

θ	ρ	velocidad	aceleración	ρ circular	R	flecha
0	740573637000	0,000000000000	0,000011242177	740573637000	778547186500	0
1	740573637012	0,011690602911	0,000011242177	740573637006	778547186506	6
2	740573637036	0,023381205818	0,000011242177	740573637024	778547186512	12
3	740573637073	0,035071808716	0,000011242177	740573637055	778547186518	18
4	740573637122	0,046762411603	0,000011242177	740573637097	778547186524	24

θ	ρ	velocidad	aceleración	ρ circular	R	flecha
1680	740590801625	19,636952092897	0,000011236575	740590796516	778547191609	5109
1681	740590822057	19,648636869762	0,000011236568	740590816949	778547191609	5109
1682	740590842502	19,660321639692	0,000011236561	740590837393	778547191609	5109
1683	740590862958	19,672006402683	0,000011236555	740590857850	778547191609	5109
1684	740590883427	19,683691158731	0,000011236548	740590878319	778547191609	5109

θ	ρ	velocidad	aceleración	ρ circular	R	flecha
3367	740642544062	39,336015828293	0,000011219688	740642544052	778547186510	10
3368	740642584979	39,347683045104	0,000011219675	740642584975	778547186504	4
3369	740642625909	39,359350248029	0,000011219661	740642625910	778547186498	-2
3370	740642666850	39,371017437066	0,000011219648	740642666858	778547186492	-8
3371	740642707804	39,382684612210	0,000011219635	740642707817	778547186486	-14

θ	ρ	velocidad	aceleración	ρ circular	R	flecha
51474	755477131320	513,829876703910	0,000006520479	755478935921	778545380502	-1805998
51475	755477665653	513,836657270204	0,000006520314	755479470254	778545380502	-1805998
51476	755478199992	513,843437665650	0,000006520150	755480004593	778545380502	-1805998
51477	755478734339	513,850217890249	0,000006519986	755480538940	778545380502	-1805998
51478	755479268692	513,856997943996	0,000006519822	755481073293	778545380502	-1805998

θ	ρ	velocidad	aceleración	ρ circular	R	flecha
82896	774767873592	635,626850623363	0,000000808757	774767873760	778547186332	-168
82897	774768534574	635,627691640373	0,000000808569	774768534644	778547186430	-70
82898	774769195556	635,628532461439	0,000000808381	774769195529	778547186527	27
82899	774769856540	635,629373086560	0,000000808192	774769856415	778547186624	124
82900	774770517524	635,630213515737	0,000000808004	774770517302	778547186722	222

θ	ρ	velocidad	aceleración	ρ circular	R	flecha
87202	777619458878	637,436206129261	0,000000000310	777619035799	778547610083	423583
87203	777620121740	637,436206451315	0,000000000123	777619698562	778547610182	423682
87204	777620784602	637,436206579021	-0,000000000064	777620361326	778547610281	423781
87205	777621447464	637,436206512379	-0,000000000251	777621024089	778547610379	423879
87206	777622110326	637,436206251390	-0,000000000438	777621686852	778547610478	423978

θ	ρ	velocidad	aceleración	ρ circular	R	flecha
122017	799330934759	526,349585369819	-0,000005874081	799328384951	778549738379	2551879
122018	799331482097	526,343476984362	-0,000005874223	799328932289	778549738379	2551879
122019	799332029429	526,337368451152	-0,000005874365	799329479621	778549738379	2551879
122020	799332576754	526,331259770192	-0,000005874508	799330026946	778549738379	2551879
122021	799333124073	526,325150941483	-0,000005874650	799330574265	778549738379	2551879

θ	ρ	velocidad	aceleración	ρ circular	R	flecha
179998	816520735989	0,021206448125	-0,000010196508	816520735978	778547186511	11
179999	816520736000	0,010603224064	-0,000010196508	816520735994	778547186506	6
180000	816520736000	0,000000000000	-0,000010196508	816520736000	778547186500	0
180001	816520735989	-0,010603224064	-0,000010196508	816520735994	778547186495	-5
180002	816520735967	-0,021206448125	-0,000010196508	816520735978	778547186489	-11

θ	ρ	velocidad	aceleración	ρ circular	R	flecha
236301	800241919937	-515,868007959498	-0,000006110163	800239912077	778549195927	2009427
236302	800241383486	-515,874361843506	-0,000006110024	800239375626	778549195927	2009427
236303	800240847028	-515,880715583195	-0,000006109885	800238839168	778549195927	2009427
236304	800240310563	-515,887069178561	-0,000006109747	800238302703	778549195927	2009427
236305	800239774092	-515,893422629603	-0,000006109608	800237766232	778549195927	2009427

θ	ρ	velocidad	aceleración	ρ circular	R	flecha
270363	779232381387	-636,862337505718	-0,000000453042	779232381208	778547186679	179
270364	779231719121	-636,862808618856	-0,000000452857	779231719040	778547186580	80
270365	779231056854	-636,863279539023	-0,000000452671	779231056872	778547186482	-18
270366	779230394588	-636,863750266218	-0,000000452486	779230394704	778547186384	-116
270367	779229732320	-636,864220800442	-0,000000452300	779229732535	778547186285	-215

θ	ρ	velocidad	aceleración	ρ circular	R	flecha
272794	777621447464	-637,436206251389	-0,000000000251	777621686852	778546946826	-239674
272795	777620784602	-637,436206512378	-0,000000000064	777621024089	778546946727	-239773
272796	777620121740	-637,436206579020	0,000000000123	777620361326	778546946629	-239871
272797	777619458878	-637,436206451314	0,000000000310	777619698562	778546946530	-239970
272798	777618796015	-637,436206129260	0,000000000497	777619035799	778546946431	-240069

θ	ρ	velocidad	aceleración	ρ circular	R	flecha
306833	756391164867	-525,063172040391	0,000006239946	756393509712	778544839760	-2346740
306834	756390618867	-525,056683196640	0,000006240113	756392963712	778544839760	-2346740
306835	756390072874	-525,050194178943	0,000006240281	756392417719	778544839760	-2346740
306836	756389526887	-525,043704987303	0,000006240448	756391871733	778544839760	-2346740
306837	756388980908	-525,037215621721	0,000006240615	756391325753	778544839760	-2346740

θ	ρ	velocidad	aceleración	ρ circular	R	flecha
359998	740573637012	-0,023381205813	0,000011242177	740573637024	778547186488	-12
359999	740573637000	-0,011690602907	0,000011242177	740573637006	778547186494	-6
360000	740573637000	0,000000000004	0,000011242177	740573637000	778547186500	0
360001	740573637012	0,011690602915	0,000011242177	740573637006	778547186506	6
360002	740573637036	0,023381205822	0,000011242177	740573637024	778547186512	12

6.5.7 Saturn's orbit

Órbita Saturno	
Afelio, A	1513325781000
Perihelio, P	1353572955000
Radio, R	1433449368000
Descentrado del Sol, D	79876413000
Periodo, T	929639263
w	6,75873487410954E-009
E	1431222152311,0400
k	670391622158453000000
incremento de tiempo	2582,331286
factor	0,9996116202
velocidad cte. órbita circular	9688,3042
flecha máxima	6874181
flecha mínima	-6230878

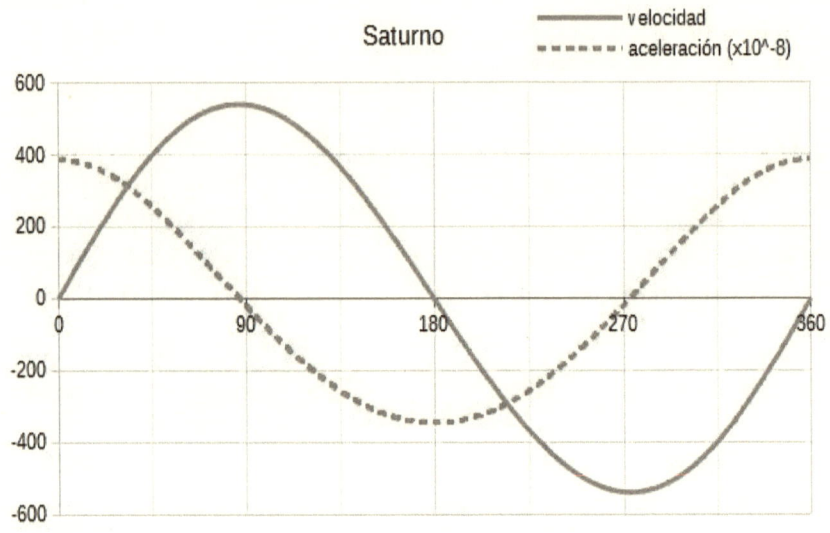

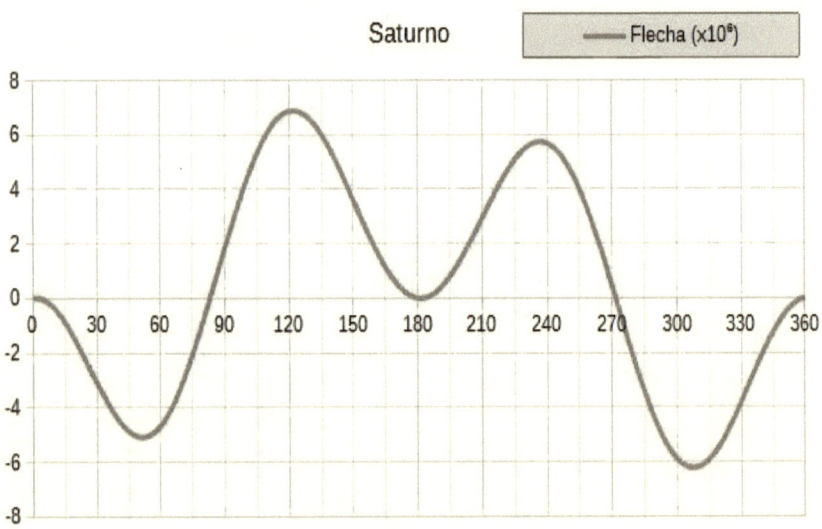

θ	ρ	velocidad	aceleración	ρ circular	R	flecha
0	1353572955000	0,000000000000	0,000003862614	1353572955000	1433449368000	0
1	1353572955026	0,009974550127	0,000003862614	1353572955013	1433449368013	13
2	1353572955077	0,019949100250	0,000003862614	1353572955052	1433449368026	26
3	1353572955155	0,029923650366	0,000003862614	1353572955116	1433449368039	39
4	1353572955258	0,039898200472	0,000003862614	1353572955206	1433449368051	51

θ	ρ	velocidad	aceleración	ρ circular	R	flecha
1286	1353594269404	12,825996836244	0,000003861462	1353594261115	1433449376289	8289
1287	1353594302551	12,835968410675	0,000003861460	1353594294262	1433449376289	8289
1288	1353594335723	12,845939980479	0,000003861459	1353594327434	1433449376289	8289
1289	1353594368922	12,855911545652	0,000003861457	1353594360633	1433449376289	8289
1290	1353594402146	12,865883106191	0,000003861455	1353594393857	1433449376289	8289

θ	ρ	velocidad	aceleración	ρ circular	R	flecha
2576	1353658431935	25,684197814007	0,000003857994	1353658431913	1433449368022	22
2577	1353658498286	25,694160431582	0,000003857990	1353658498277	1433449368009	9
2578	1353658564662	25,704123039896	0,000003857986	1353658564666	1433449367996	-4
2579	1353658631065	25,714085638944	0,000003857983	1353658631081	1433449367983	-17
2580	1353658697493	25,724048228723	0,000003857979	1353658697522	1433449367970	-30

θ	ρ	velocidad	aceleración	ρ circular	R	flecha
51342	1384953148322	436,330800256474	0,000002223280	1384958236452	1433444274701	-5093299
51343	1384954275087	436,336541501845	0,000002223223	1384959363218	1433444274701	-5093299
51344	1384955401868	436,342282600358	0,000002223166	1384960489998	1433444274701	-5093299
51345	1384956528663	436,348023552011	0,000002223109	1384961616793	1433444274701	-5093299
51346	1384957655473	436,353764356803	0,000002223052	1384962743603	1433444274701	-5093299

θ	ρ	velocidad	aceleración	ρ circular	R	flecha
83158	1426140397033	538,869499600482	0,000000233652	1426140397375	1433449367658	-342
83159	1426141788574	538,870102966668	0,000000233587	1426141788647	1433449367927	-73
83160	1426143180117	538,870706166750	0,000000233523	1426143179922	1433449368195	195
83161	1426144571661	538,871309200729	0,000000233459	1426144571198	1433449368464	464
83162	1426145963207	538,871912068604	0,000000233394	1426145962475	1433449368733	733

θ	ρ	velocidad	aceleración	ρ circular	R	flecha
86803	1431219456375	539,968435627793	0,000000000123	1431218471656	1433450354251	986251
86804	1431220850752	539,968435946183	0,000000000060	1431219865762	1433450354523	986523
86805	1431222245130	539,968436099897	-0,000000000004	1431221259869	1433450354794	986794
86806	1431223639507	539,968436088935	-0,000000000068	1431222653975	1433450355066	987066
86807	1431225033884	539,968435913298	-0,000000000132	1431224048081	1433450355337	987337

θ	ρ	velocidad	aceleración	ρ circular	R	flecha
121856	1477172768924	444,979771262305	-0,000002004456	1477165901997	1433456242181	6874181
121857	1477173917996	444,974595091830	-0,000002004504	1477167051069	1433456242181	6874181
121858	1477175067054	444,969418797924	-0,000002004552	1477168200127	1433456242181	6874181
121859	1477176216099	444,964242380588	-0,000002004600	1477169349172	1433456242181	6874181
121860	1477177365131	444,959065839826	-0,000002004648	1477170498204	1433456242181	6874181

θ	ρ	velocidad	aceleración	ρ circular	R	flecha
179998	1513325780977	0,017843192070	-0,000003454861	1513325780954	1433449368024	24
179999	1513325781000	0,008921596036	-0,000003454861	1513325780988	1433449368012	12
180000	1513325781000	-0,000000000001	-0,000003454861	1513325781000	1433449368000	0
180001	1513325780977	-0,008921596038	-0,000003454861	1513325780988	1433449367989	-11
180002	1513325780931	-0,017843192072	-0,000003454861	1513325780954	1433449367977	-23

θ	ρ	velocidad	aceleración	ρ circular	R	flecha
236855	1478641581826	-438,205842247795	-0,000002065464	1478635855196	1433455100497	5732497
236856	1478640450220	-438,211175960052	-0,000002065417	1478634723590	1433455100497	5732497
236857	1478639318600	-438,216509551115	-0,000002065370	1478633591969	1433455100498	5732498
236858	1478638186966	-438,221843020984	-0,000002065323	1478632460336	1433455100498	5732497
236859	1478637055318	-438,227176369656	-0,000002065276	1478631328688	1433455100497	5732497

θ	ρ	velocidad	aceleración	ρ circular	R	flecha
271683	1433328905103	-539,780595270552	-0,000000096138	1433328904681	1433449368423	423
271684	1433327511210	-539,780843529473	-0,000000096074	1433327511058	1433449368152	152
271685	1433326117316	-539,781091624500	-0,000000096011	1433326117435	1433449367881	-119
271686	1433324723422	-539,781339555632	-0,000000095947	1433324723812	1433449367610	-390
271687	1433323329527	-539,781587322869	-0,000000095884	1433323330187	1433449367339	-661

θ	ρ	velocidad	aceleración	ρ circular	R	flecha
273193	1431223639507	-539,968435913300	-0,000000000068	1431224048081	1433448958790	-409210
273194	1431222245130	-539,968436088937	-0,000000000004	1431222653975	1433448958518	-409482
273195	1431220850752	-539,968436099899	0,000000000060	1431221259869	1433448958247	-409753
273196	1431219456375	-539,968435946185	0,000000000123	1431219865762	1433448957976	-410025
273197	1431218061997	-539,968435627794	0,000000000187	1431218471656	1433448957704	-410296

θ	ρ	velocidad	aceleración	ρ circular	R	flecha
307359	1386428030905	-443,664302756569	0,000002148958	1386434255246	1433443137122	-6230878
307360	1386426885231	-443,658753435257	0,000002149016	1386433109573	1433443137122	-6230878
307361	1386425739571	-443,653203965099	0,000002149073	1386431963913	1433443137122	-6230878
307362	1386424593926	-443,647654346094	0,000002149131	1386430818268	1433443137122	-6230878
307363	1386423448295	-443,642104578246	0,000002149188	1386429672637	1433443137122	-6230878

θ	ρ	velocidad	aceleración	ρ circular	R	flecha
359998	1353572955026	-0,019949100249	0,000003862614	1353572955052	1433449367974	-26
359999	1353572955000	-0,009974550126	0,000003862614	1353572955013	1433449367987	-13
360000	1353572955000	0,000000000001	0,000003862614	1353572955000	1433449368000	0
360001	1353572955026	0,009974550128	0,000003862614	1353572955013	1433449368013	13
360002	1353572955077	0,019949100251	0,000003862614	1353572955052	1433449368026	26

6.5.8 Orbit of Uranus

Órbita Urano	
Afelio, A	3004419704000
Perihelio, P	2748938461000
Radio, R	2876679082500
Descentrado del Sol, D	127740621500
Periodo, T	2661041808
w	2,36117496849925E-009
E	2873841484374,5000
k	66211934696482700000
incremento de tiempo	7391,782800
factor	0,9997534192
velocidad cte. órbita circular	6792,3426
flecha máxima	7245350
flecha mínima	-6724439

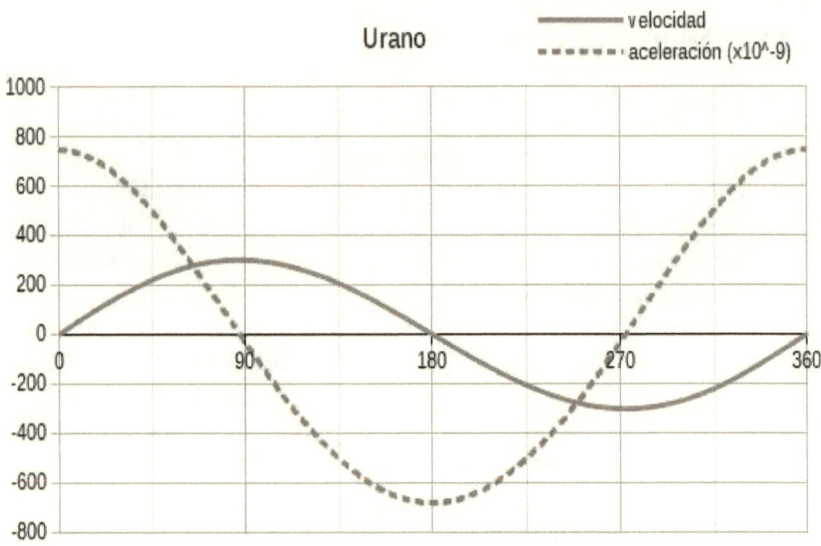

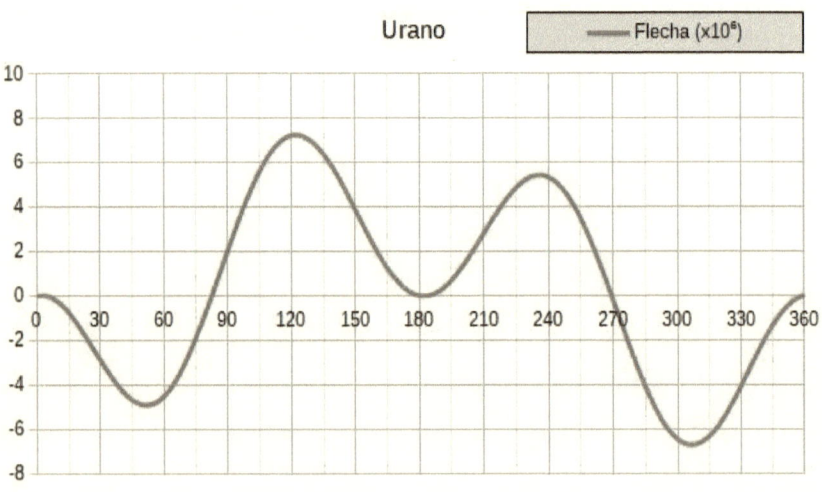

θ	ρ	velocidad	aceleración	ρ circular	R	flecha
0	2748938461000	0,000000000000	0,000000745083	2748938461000	2876679082500	0
1	2748938461041	0,005507491795	0,000000745083	2748938461020	2876679082520	20
2	2748938461122	0,011014983587	0,000000745083	2748938461081	2876679082541	41
3	2748938461244	0,016522475376	0,000000745083	2748938461183	2876679082561	61
4	2748938461407	0,022029967159	0,000000745083	2748938461326	2876679082581	81

θ	ρ	velocidad	aceleración	ρ circular	R	flecha
2028	2749022208372	11,166526912291	0,000000744549	2749022187726	2876679103146	20646
2029	2749022290953	11,172030457952	0,000000744549	2749022270307	2876679103146	20646
2030	2749022373575	11,177533999722	0,000000744548	2749022352929	2876679103146	20646
2031	2749022456238	11,183037537599	0,000000744548	2749022435591	2876679103146	20646
2032	2749022538941	11,188541071582	0,000000744547	2749022518295	2876679103146	20646

θ	ρ	velocidad	aceleración	ρ circular	R	flecha
4068	2749275230722	22,382961588194	0,000000742937	2749275230693	2876679082529	29
4069	2749275396213	22,388453213762	0,000000742935	2749275396204	2876679082509	9
4070	2749275561744	22,393944831534	0,000000742934	2749275561755	2876679082488	-12
4071	2749275727316	22,399436441510	0,000000742933	2749275727347	2876679082468	-32
4072	2749275892928	22,404928043686	0,000000742932	2749275892980	2876679082448	-52

θ	ρ	velocidad	aceleración	ρ circular	R	flecha
51502	2798943276260	242,609229772446	0,000000434801	2798948178699	2876674176930	-4905570
51503	2798945069598	242,612443724511	0,000000434790	2798949972038	2876674176930	-4905570
51504	2798946862960	242,615657596520	0,000000434779	2798951765400	2876674176930	-4905570
51505	2798948656346	242,618871388471	0,000000434768	2798953558786	2876674176930	-4905570
51506	2798950449756	242,622085100364	0,000000434757	2798955352196	2876674176930	-4905570

θ	ρ	velocidad	aceleración	ρ circular	R	flecha
82528	2862870535921	300,535519871260	0,000000061562	2862870536277	2876679082143	-357
82529	2862872757417	300,535974926020	0,000000061550	2862872757504	2876679082413	-87
82530	2862874978918	300,536429888107	0,000000061537	2862874978734	2876679082684	184
82531	2862877200421	300,536884757520	0,000000061525	2862877199967	2876679082954	454
82532	2862879421928	300,537339534261	0,000000061512	2862879421204	2876679083224	724

θ	ρ	velocidad	aceleración	ρ circular	R	flecha
87453	2873838572649	301,655167347002	0,000000000016	2873837225103	2876680431377	1348877
87454	2873840802418	301,655167467084	0,000000000004	2873839454597	2876680431652	1349152
87455	2873843032188	301,655167495208	-0,000000000009	2873841684092	2876680431927	1349427
87456	2873845261957	301,655167431375	-0,000000000021	2873843913586	2876680432203	1349703
87457	2873847491727	301,655167275585	-0,000000000034	2873846143080	2876680432478	1349978

θ	ρ	velocidad	aceleración	ρ circular	R	flecha
122072	2946503580440	249,530841644115	-0,000000390529	2946496339981	2876686327850	7245350
122073	2946505424897	249,527954938053	-0,000000390539	2946498184437	2876686327850	7245350
122074	2946507269332	249,525068161414	-0,000000390548	2946500028872	2876686327850	7245350
122075	2946509113745	249,522181314198	-0,000000390558	2946501873286	2876686327850	7245350
122076	2946510958138	249,519294396408	-0,000000390567	2946503717678	2876686327850	7245350

θ	ρ	velocidad	aceleración	ρ circular	R	flecha
179998	3004419703963	0,010078322942	-0,000000681725	3004419703925	2876679082538	38
179999	3004419704001	0,005039161472	-0,000000681725	3004419703981	2876679082519	19
180000	3004419704001	0,000000000000	-0,000000681725	3004419704000	2876679082501	1
180001	3004419703963	-0,005039161472	-0,000000681725	3004419703981	2876679082482	-18
180002	3004419703889	-0,010078322942	-0,000000681725	3004419703925	2876679082463	-37

θ	ρ	velocidad	aceleración	ρ circular	R	flecha
235898	2950201360532	-243,526811545071	-0,000000409635	2950195942786	2876684503732	5421232
235899	2950199560412	-243,529839478443	-0,000000409626	2950194142666	2876684503732	5421232
235900	2950197760270	-243,532867343194	-0,000000409616	2950192342524	2876684503732	5421232
235901	2950195960106	-243,535895139322	-0,000000409607	2950190542360	2876684503732	5421232
235902	2950194159919	-243,538922866827	-0,000000409598	2950188742173	2876684503732	5421232

θ	ρ	velocidad	aceleración	ρ circular	R	flecha
269331	2881003529035	-301,181464537387	-0,000000039810	2881003528566	2876679082970	470
269332	2881001302765	-301,181758805315	-0,000000039798	2881001302569	2876679082697	197
269333	2880999076493	-301,182052982114	-0,000000039785	2880999076569	2876679082424	-76
269334	2880996850219	-301,182347067782	-0,000000039773	2880996850568	2876679082151	-350
269335	2880994623942	-301,182641062319	-0,000000039761	2880994624564	2876679081877	-623

θ	ρ	velocidad	aceleración	ρ circular	R	flecha
272543	2873845261957	-301,655167275585	-0,000000000021	2873846143080	2876678200507	-881993
272544	2873843032188	-301,655167431375	-0,000000000009	2873843913586	2876678200231	-882269
272545	2873840802418	-301,655167495208	0,000000000004	2873841684092	2876678199956	-882544
272546	2873838572649	-301,655167467084	0,000000000016	2873839454597	2876678199681	-882819
272547	2873836342879	-301,655167347002	0,000000000029	2873837225103	2876678199405	-883095

θ	ρ	velocidad	aceleración	ρ circular	R	flecha
306455	2802653934847	-249,007081063682	0,000000412436	2802660654766	2876672358061	-6724439
306456	2802652094263	-249,004032430021	0,000000412447	2802658814182	2876672358061	-6724439
306457	2802650253702	-249,000983714521	0,000000412458	2802656973621	2876672358061	-6724439
306458	2802648413163	-248,997934917183	0,000000412469	2802655133083	2876672358061	-6724439
306459	2802646572647	-248,994886038006	0,000000412480	2802653292567	2876672358061	-6724439

θ	ρ	velocidad	aceleración	ρ circular	R	flecha
359998	2748938461041	-0,011014983587	0,000000745083	2748938461081	2876679082459	-41
359999	2748938461000	-0,005507491795	0,000000745083	2748938461020	2876679082480	-20
360000	2748938461000	0,000000000000	0,000000745083	2748938461000	2876679082500	0
360001	2748938461041	0,005507491795	0,000000745083	2748938461020	2876679082520	20
360002	2748938461122	0,011014983587	0,000000745083	2748938461081	2876679082541	41

6.5.9 Orbit of Neptun

Órbita Neptuno	
Afelio, A	4553946490000
Perihelio, P	4452940833000
Radio, R	4503443661500
Descentrado del Sol, D	50502828500
Periodo, T	5200416000
w	1,2082082101085E-009
E	4503160476445,1900
k	66654016401529300000
incremento de tiempo	14445,600000
factor	0,9999842796
velocidad cte. órbita circular	5441,0976
flecha máxima	539054
flecha mínima	-535755

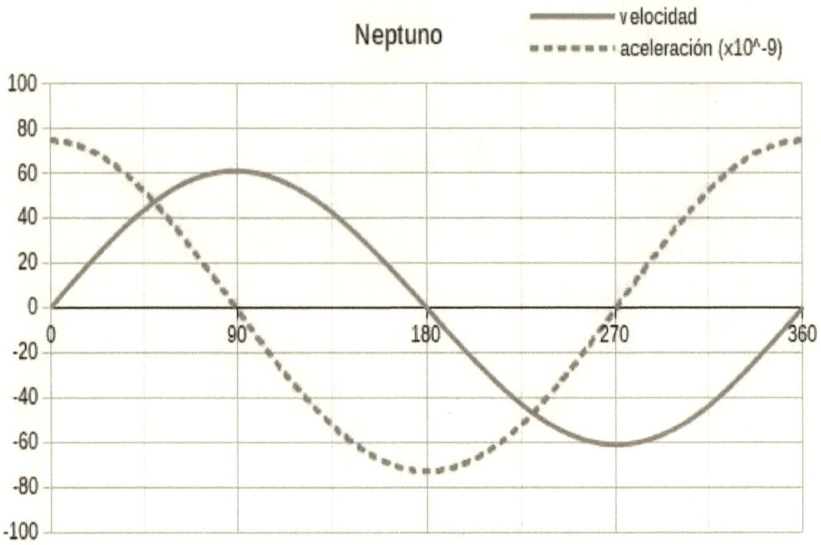

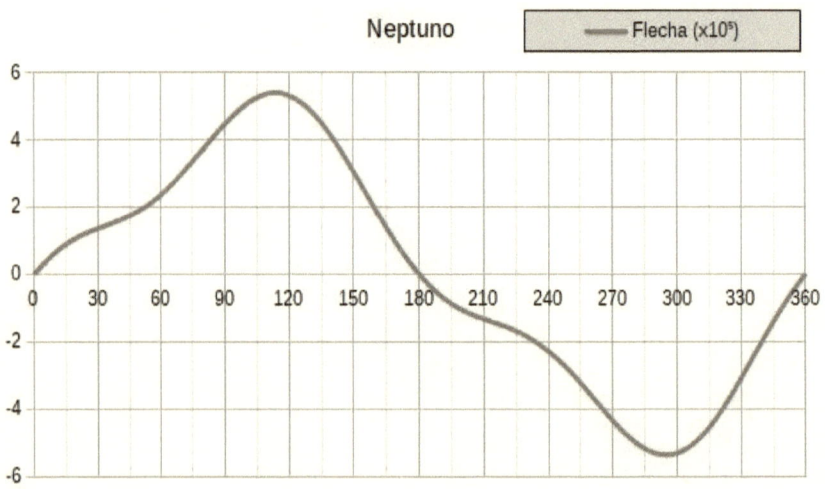

θ	ρ	velocidad	aceleración	ρ circular	R	flecha
0	4452940833000	0,000000000000	0,000000074557	4452940833000	4503443661500	0
1	4452940833016	0,001077025124	0,000000074557	4452940833008	4503443661508	8
2	4452940833047	0,002154050248	0,000000074557	4452940833031	4503443661516	16
3	4452940833093	0,003231075372	0,000000074557	4452940833070	4503443661523	23
4	4452940833156	0,004308100494	0,000000074557	4452940833124	4503443661531	31

θ	ρ	velocidad	aceleración	ρ circular	R	flecha
89355	4503158753598	61,018411604307	0,000000000003	4503158311407	4503444103719	442219
89356	4503159635045	61,018411640639	0,000000000001	4503159192848	4503444103726	442225
89357	4503160516493	61,018411658383	0,000000000000	4503160074288	4503444103732	442232
89358	4503161397940	61,018411657538	-0,000000000001	4503160955729	4503444103739	442239
89359	4503162279388	61,018411638105	-0,000000000003	4503161837169	4503444103746	442246

θ	ρ	velocidad	aceleración	ρ circular	R	flecha
113578	4523881987921	55,671500911018	-0,000000030042	4523881448895	4503444200554	539054
113579	4523882792123	55,671066929292	-0,000000030044	4523882253097	4503444200554	539054
113580	4523883596318	55,670632930838	-0,000000030045	4523883057293	4503444200554	539054
113581	4523884400508	55,670198915658	-0,000000030046	4523883861482	4503444200554	539054
113582	4523885204691	55,669764883750	-0,000000030047	4523884665665	4503444200554	539054

θ	ρ	velocidad	aceleración	ρ circular	R	flecha
179998	4553946489985	0,002106273808	-0,000000072904	4553946489970	4503443661515	15
179999	4553946490000	0,001053136894	-0,000000072904	4553946489992	4503443661508	8
180000	4553946490000	-0,000000000020	-0,000000072904	4553946490000	4503443661500	0
180001	4553946489985	-0,001053136935	-0,000000072904	4553946489992	4503443661492	-8
180002	4553946489954	-0,002106273849	-0,000000072904	4553946489970	4503443661485	-15

θ	ρ	velocidad	aceleración	ρ circular	R	flecha
270641	4503161397940	-61,018411638107	-0,000000000001	4503161837169	4503443222243	-439257
270642	4503160516493	-61,018411657540	0,000000000000	4503160955729	4503443222236	-439264
270643	4503159635045	-61,018411658384	0,000000000001	4503160074288	4503443222229	-439271
270644	4503158753598	-61,018411640641	0,000000000003	4503159192848	4503443222222	-439278
270645	4503157872150	-61,018411604309	0,000000000004	4503158311407	4503443222216	-439284

θ	ρ	velocidad	aceleración	ρ circular	R	flecha
294717	4482560804837	-55,686155225060	0,000000030280	4482561340564	4503443125745	-535755
294718	4482560000423	-55,685717815851	0,000000030281	4482560536151	4503443125745	-535755
294719	4482559196016	-55,685280389442	0,000000030282	4482559731743	4503443125745	-535755
294720	4482558391615	-55,684842945835	0,000000030283	4482558927342	4503443125745	-535755
294721	4482557587220	-55,684405485030	0,000000030285	4482558122948	4503443125745	-535755

θ	ρ	velocidad	aceleración	ρ circular	R	flecha
359998	4452940833016	-0,002154050203	0,000000074557	4452940833031	4503443661484	-16
359999	4452940833000	-0,001077025079	0,000000074557	4452940833008	4503443661492	-8
360000	4452940833000	0,000000000045	0,000000074557	4452940833000	4503443661500	0
360001	4452940833016	0,001077025170	0,000000074557	4452940833008	4503443661508	8
360002	4452940833047	0,002154050294	0,000000074557	4452940833031	4503443661516	16

7. Proposed calculation of the equilibrium distance

In the calculation of the orbits, at the equilibrium distance d_0 we have called it E and its value is given by the geometric mean of the aphelion and perihelion for the planets or the apogee and perigee for the case of the Moon. This value has been obtained by adjusting the orbits so that when a time equal to the sidereal period T passes, the star returns to the eplanet in the same position and at the same distance from the Sun or Earth in the case of the Moon. In the worksheet of the orbits, this is the only value that synchronizes the annual movement of translation with the oscillatory movement of approach and distance.

Next we will try to obtain the value of d_0 from the law of orbital gravitation.

$$F = GMm\left(\frac{1}{d^2} - \frac{1}{d_0^2}\right)$$

It has been seen previously that the equilibrium distance between two bodies of radius and mass R, M and r, m respectively, is that in which the body of least mass m, receives from the body of greater mass M, a flow equal to double the one that emits in the direction that unites M and m. It is clear from this definition that d0 is directly proportional to M and a r and inversely proportional to m and R. It therefore remains:

$$d_0 = Cte \frac{r^2}{R} \frac{M}{m}$$

This is the simplest formula that is homogeneous. Its dimensions are [L] 1 in both members, which corresponds to a distance. The values of d_0, r and R are known, but the values of the constant and those of M and m are unknown since the masses that come in the tables have been calculated with a law that is not applicable, so if you substitute values will not match the value of d_0 with the one calculated by this formula. Assuming the value of the constant equal to 1, by this method can only calculate the relationship between M and m since there is an equation and two unknowns. It is going to be assumed that the mass of the Sun is correct and from it and from the formula deduced the rest of the masses will be calculated so that they fulfill that relation.

Clearing in the formula results:

$$m = \frac{r^2}{R} \frac{M}{d_0}$$

M and R are referred to the Sun for the planets and the Earth for the Moon. In the following table, we compare the masses and densities currently accepted with the new ones calculated by this formula. The radii and the equilibrium distances are expressed in km, the mass in kg and the densities in kg/m³.

	radio	distancia de equilibrio d_0	Old data		New data	
			masa	densidad	masa	densidad
Mercurio	2440	56671508	3,30E+23	5430	3,00E+23	4938,2
Venus	6052	108205635	4,87E+24	5240	9,68E+23	1042,6
Tierra	6371	149577371	5,97E+24	5515	7,76E+23	716,5
Marte	3397	226944568	6,42E+23	3934	1,45E+23	885,6
Ceres	476	412824154	9,43E+20	2020	1,57E+21	3473,1
Júpiter	71492	306085281	1,90E+27	1336	4,78E+25	31,2
Saturno	60268	1431222152	5,69E+26	690	7,26E+24	7,9
Urano	25559	2873841484	8,69E+25	1274	6,50E+23	9,3
Neptuno	24786	4503160476	1,02E+26	1640	3,90E+23	6,1
Luna	1737	380179	7,35E+22	3340	9,67E+20	44,0

Table of corrected densities and masses

This table contains several surprises. Except for Ceres, the mass is smaller than what is currently admitted. For Mercury the difference is small, for the Earth the mass is almost 7.7 times smaller and for the gas giant planets, the mass is of the order of one hundred times smaller and the densities clearly show that they are gaseous. The Moon results with a mass 100 times less than expected and a density of 44.0 kg / m³ that is impossible unless it is hollow, rather than practically an empty shell. The Earth with a density of 716.5 kg / m³ must also have a hollow eplanet since the average density in the crust is of the order of 2000 kg / m³.

Let's look at the thickness assuming that these new data are true. For the Moon, assuming a density of the crust of 2000 kg / m³, if x is the thickness of the crust in meters, we have:

$$9,667\,E\,20 = \frac{4\pi}{3}(1737000^3 - (1737000 - x)^3)\,2000$$

whose solution is x = 12843 m.

That is, supposing that the previous data are correct and that the crust of the Moon had a uniform density of 2000 kg / (m^3), then the Moon would be a hollow shell whose thickness would not reach 13 km.

For the Earth, assuming the same average density for the crust, we would have:

$$7,76\,E\,23 = \frac{4\pi}{3}(6371000^3 - (6371000-x)^3)\,2000$$

whose solution is x = 875484 m, that is, if this is true, the Earth must also have a hollow eplanet and it does not reach 900 km thick.

It must be said that as long as the value of the constant and the exact form of the relationship are not known, these calculations show nothing. It is not our intention to enter into old polemics. The mass of the Sun has been considered to be the same as the tables drawn with Newton's law, which does not stop being an assumption until the correct data is available. The value of the constant could also have been adjusted so that the Earth had the currently admitted density of 5515 kg / m³ but that would imply an impossible density for Mercury, so the constant has been left at 1. Nor have they been considered, apart from Ceres , the objects of greater mass of the asteroid belt such as Palas, Vesta and others that having a smaller radius than Ceres and a similar equilibrium distance, must have a higher density so that they are at a distance than they are, which could lead one to think that These asteroids are mainly composed of heavy metals such as iron, nickel, gold, platinum or tungsten for example.

Here the interesting thing is to see that the ordering of the orbits of the planets is not random. This arrangement responds to a relationship between its mass and its radius. A planet with great mass emits many gravitons and with great volume receives many. Its equilibrium distance from the Sun depends on the relationship between both variables. The logical thing is that the heavier and denser planets are closer and the more voluminous and gaseous are further away. The case of Ceres is also logical, it is lower radius but it is compensated with a higher density. Until the data is revised with the new formula, we can not speak of absolute values of mass, only of relationships.

8. Impact of asteroids on Earth

This topic has been left to the end, which is of vital importance because what is at stake is the survival of many species on Earth, including humans.

There are ongoing studies and projects to monitor large objects that approach the Earth and divert or destroy those that have a collision course. These projects cost millions of dollars. The expense was justified if it were to save the Earth. The fact is that the shot can come out by the butt and the opposite effect to the wished one takes place.

In the study of the impact probability of an asteroid on Earth there is now a new variable that is the repulsive term of the gravitational law that was not taken into account with Newton's law. Gravitational force becomes repulsive at distances less than equilibrium. The equilibrium distance depends on the relationship between the masses and radii of the objects. The situation is the following one: for a asteroid of great mass is very high the probability that it is repelled by the terrestrial gravity diverting its trajectory as it approaches. On the contrary, a fragment of that asteroid could be attracted by having smaller mass. After an explosion the center of gravity of the fragments does not deviate. The atmosphere would protect us from small fragments but not from intermediate ones. In short, fragmenting an asteroid without taking into account the mass ratio and the equilibrium distance is a true suicide of ignorant people. We planted causing just what we want to avoid.

It turns out that wise nature with the repulsive term of gravitation protects us from objects of great mass and the atmosphere protects us from objects of small mass. The size here is very important. There is an intermediate mass that is dangerous. There is evidence that on Earth these objects of intermediate mass have impacted in the past. As we have seen in previous chapters, it is not possible to determine the critical mass until we know the true value of the gravitational constant and the mass of the Earth. As stated in the introduction, we look forward to the help and collaboration of the scientific community for its early determination.

9. Annex: Spreadsheet formulas

Spreadsheet formulas LibreOffice Version: 6.2.4.2 (x64) used:

Only Mercury is put because the calculations are the same for all planets. Data entries vary but not structure or formulas.

	B	C
	\multicolumn{2}{c}{Órbita Mercurio}	
3	Afelio, A	69816877400
4	Perihelio, P	46001195600
5	Radio, R	57909036500
6	Descentrado, D	11907840900
7	Periodo, T	7600428
8	w	0,0000008266883532
9	E	56671508127,6175
10	k	63213202493004700000
11	tiempo	21,1123
12	factor	0,9946682858
13	velocidad orbital cte.	47872,7260214459
14	Flecha máxima	15077140
15	Flecha mínima	-10115734

9.1 Mercury orbit formulas

C3: Input data, Afelion

C4: Input data, Perihelion

C5=(C3+C4)/2

C6=(C3-C4)/2

C7: Input data, sidereal period

C8=2*PI()/C7

C9=RAIZ(C3*C4).

C10=C8^2*C5*C9^2*C12/2

C11=C7/360000

C12: Adjustment parameter. (Experiment until perihelion velocity = 0 when a full revolution is given ie for θ = 360000 in this case).

C13=C8*C5

C14=MAX(R4:R360004).

C15=MIN(R4:R360004).

	J	K	L	M	N	O	P
	θ	ρ	velocidad	aceleración	ρ circular	R	Flecha
4	0	46001195600	0,00000000	0,010189952333	46001195600	57909036500	0,0000
5	1	46001195605	0,21513333	0,010189952327	46001195602	57909036502	2,2588

J4=0; (begins to count the angle in the perihelion)

J5=1; (drag row J down to increase by one unit to J360004 = 360000).

K4=C4; (begins in the perihelion).

K5=M4+N5*$C$11

K6=M5+N6*$C$11; (Row K is dragged down to K360004).

L4=0; (velocity equal to zero in the perihelion).

L5=N4+O4*$C$11

L6=N5+O5*$C$11; (Row L is dragged down to L360004).

M4=$C$10*(1/(M4^2)-1/($C$9^2)).

M5=$C$10*(1/(M5^2)-1/($C$9^2)); (Row M is dragged down to M360004).

N4=RAIZ($C$5^2+$C$6^2-2*$C$5*$C$6*COS(L4*PI()/180000))

and dragging...

O4=$C$6*COS(L4*PI()/180000)+RAIZ(M4^2-$C$6^2*(SENO(L4*PI()/180000))^2);

P4=O4-$C$5; .also dragged like the previous ones. (This is the arrow "flecha")

END

Celestial orbits

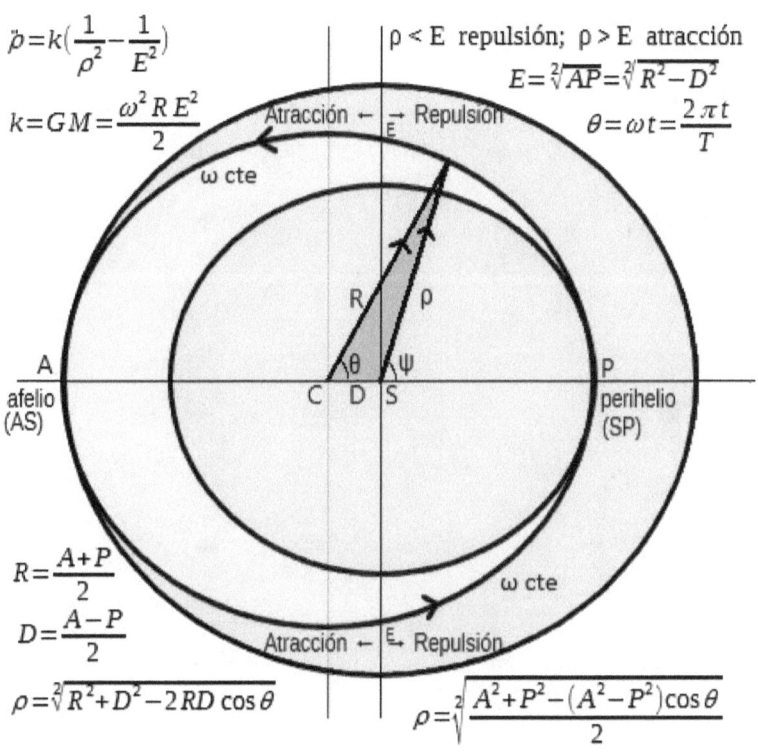

www.ingramcontent.com/pod-product-compliance
Lightning Source LLC
Chambersburg PA
CBHW030014190526
45157CB00016B/2710